Topics in

Noncommutative Geometry

M. B. PORTER LECTURES

RICE UNIVERSITY, DEPARTMENT OF MATHEMATICS

SALOMON BOCHNER, FOUNDING EDITOR

Topics in Noncommutative Geometry

YURI I. MANIN

PRINCETON UNIVERSITY PRESS
PRINCETON, NEW JERSEY

Published by Princeton University Press, 41 William Street,
Princeton, New Jersey 08540
In the United Kingdom: Princeton University Press, Oxford

Library of Congress Cataloging-in-Publication Data
Manin, IU I.
Topics in noncommutative geometry / Yuri I. Manin.
p. cm.
"M.B. Porter lecture series."
Includes bibliographical references and index.
ISBN 0-691-08588-9
1. Geometry, Algebraic. 2. Noncommutative rings. I. Title.
QA564.M36 1991
516.3'5—dc20 90-47135

This book has been composed in Times Roman

Princeton University Press books are printed on acid-free paper, and meet the guidelines for permanence and durability of the Committee on Production Guidelines for Book Longevity of the Council on Library Resources

Printed in the United States of America by Princeton University Press, Princeton, New Jersey

10 9 8 7 6 5 4 3 2 1

CONTENTS

PREFACE

After R. Descartes, I. M. Gelfand, and A. Grothendieck, it became a truism that any commutative ring is a ring of functions on an appropriate space.

Noncommutative algebra resisted geometrization longer. A recent upsurge of activity in this domain was prompted by various developments in theoretical physics, functional analysis, and algebra.

When I was invited to give the Milton Brockett Porter Lectures at Rice University in the fall of 1989, I decided to give an introduction to some current ideas in noncommutative geometry. This book was prepared before the lectures and contains more material than could be presented orally. Still, I tried to preserve some spirit of lecture notes, especially in the first chapter, which intends to be an overview of various points of departure and basic themes. The rest of the book is more specialized. Chapters 2 and 3 are devoted to supersymmetric curves and flag spaces of supergroups, respectively. Chapter 4 develops an approach to quantum groups as symmetry objects in noncommutative geometry initiated in my Montreal lectures [Ma2].

Section 1 of Chapter 1 can be read as an introduction to the entire book. The rest of Chapter 1 gives some definitions, examples, and constructions but contains practically no proofs. It should be considered a guide for further reading.

Starting with Chapter 2, we prove most of the results. The choice of material was dictated by personal interests of the author. Exposition is based upon lecture courses and seminars I have led for several years at Moscow University and elsewhere. Partly it is taken from the papers of participants of these seminars or based upon their notes.

I want to thank many people for friendly collaboration and shared insights, especially A. A. Beilinson, V. G. Drinfeld, D. A. Leites, I. B. Penkov, A. O. Radul, I. A. Skornyakov, A. Yu. Vaintrob, A. A. Voronov, M. Wodzicki.

I would also like to thank Utrecht University and the Netherlands Mathematical Society for their hospitality during my visit in January 1988 and the Netherlands Organization for Scientific Research (ZWO) for financial support, which enabled me to write part of this book. Finally, I would like to thank a referee who suggested a number of corrections and revisions incorporated in the text.

Yuri I. Manin

Topics in

Noncommutative Geometry

CHAPTER 1

An Overview

1. Sources of Noncommutative Geometry

1.1. COMMUTATIVE GEOMETRY. The classical Euclidean geometry studies properties of some special subsets of plane and space: circles, triangles, pyramids, etc. Some of the crucial notions are those of a measure (of an angle, distance, surface, volume) and of "congruence" or equality of geometric objects.

An implicit basic object that only a century ago started to become a subject of independent geometric study is the group of motions. In fact, measures can be introduced as various motion invariants, and equality can be defined in terms of orbits of this group.

Since Descartes, this geometric picture became enriched with an essential algebraic counterpart, centered around the idea of coordinates. Subsets of space can be defined by relations between the coordinates of points, motions by means of functions. Geometry can be systematically translated into algebraic language.

Relations between coordinates are written in the form, say,

$$f_i(x_1,\dots,x_n)=0,\quad i\in I \text{ (or } f_i\neq 0\text{, or } f_i\geq 0\text{)},$$

where the f_i belong to a class $\mathcal{O}$ of functions on the basic space. This class $\mathcal{O}$ depends on the type of geometric properties one wants to study. It may consist of polynomials (algebraic geometry), complex-analytic functions (complex geometry), smooth functions (differential geometry), continuous functions (topology), and measurable functions (measure theory). In certain basic situations, the transcription of geometry into algebraic language is as neat and simple as one can possibly hope. A classical example is Gelfand's theorem, which states (in modern words) that by associating to a locally compact Hausdorff topological space its $*$-algebra of complex-valued continuous functions vanishing at infinity, we get an antiequivalence of categories. (The source category should be considered with proper continuous maps as morphisms; the essential image of the functor consists of commutative $*$-algebras and $*$-homomorphisms.)

Symmetries of a geometric object are traditionally described by its automorphism group, which often is an object of the same geometric class (a topological space, an algebraic variety, etc.). Of course, such symmetries are only a particular type of morphisms, so that Klein's Erlangen program is, in principle, subsumed by the general categorical approach. Still, automorphisms and spaces with large automorphism groups are quite special and cherished objects of study in any geometrical discipline.

Since automorphisms of an object act on any linear space naturally associated to this object (functions, (co)homology, etc.), this explains the role of representation theory.

With traditional pointwise multiplication and addition, many important classes of functions form a commutative ring. This property seems to be very crucial if we want to consider an abstract as a ring of functions.

Grothendieck's algebraic geometry, via the notion of an affine scheme, shows that there is no need, in general, to ask anything more (e.g., absence of nilpotents).

However, one can ask less.

Attempts to build geometric disciplines based upon noncommutative rings are continuing. They have led already to various important, if disparate, developments. In the following, we shall consider below some motivations and approaches, stemming from physics, functional analysis, and algebraic geometry. We make no claims of completeness at all and discuss mostly subjects that appeal to the author personally.

1.2. Physics. The basic symmetry group of Euclidean geometry has a distinctly physical origin. In fact, it is the group of motions of a rigid body, the latter notion being one of the pillars of classical physics, whose remnants can be traced as late as in Einstein's discussion of space–time in terms of rulers and clocks.

It is no surprise, then, that quantum physics supplied its own stock of basic geometries. We shall not be concerned here with already traditional Hilbert space geometry but rather with more recent developments connected with supergeometry and quantum groups.

Supergeometry is a variant of classical (differential, analytic, or algebraic) geometry in which, together with the usual pairwise commuting coordinates, one considers also anticommuting ones. The latter correspond to the internal (spinlike) degrees of freedom of fermions, elementary constituents of matter such as electrons and quarks, whereas commuting coordinates are used to describe their "external," space–time position. Since anticommuting coordinates are nilpotents, supergeometry looks like a very slight extension of the classical geometry. This does not mean at all that it is a trivial extension. It reveals a lot of very concrete new structures whose resemblance to the old ones is beautiful and fascinating. In particular, a fundamental role is played by the "supersymmetry" mixing even and odd coordinates. This notion cannot be made precise in terms of classical group theory; in fact, supersymmetry is described by Lie supergroups, or, infinitesimally, Lie superalgebras. The classification of simple Lie superalgebras, due to V. Kac ([K1]), demonstrated the existence of a quite unexpected extension of the Killing–Cartan theory and drew the attention of mathematicians to supergeometry, whose first constructions were invented by physicists Yu. Golfand, E. Lichtman, A. Volkov, J. Wess, and F. Berezin.

In this book, Chapters 2 and 3 are devoted to supergeometry. We have chosen not to explain foundations and first examples, but to develop two concrete and fairly advanced subjects: a theory of supersymmetric algebraic curves and a theory of Schubert–Bruhat decomposition of superhomogeneous flag spaces. We assume that the reader is familiar with Chapters 3 and 4 of [Ma1] containing an introduction to supergeometry. The exposition in Chapter 3 closely follows [VM].

Quantum groups could have been, but were not, invented in the same way as supergroups, i.e., as symmetry objects of certain "quantum spaces," described by noncommutative rings of functions. Actually, they originated in the work of L. D. Faddeev and his school on the quantum inverse scattering method (cf. [SkTF], [Dr1], and references therein).

An intuitive notion of a quantum space is based upon one of the schemes of quantization of classical Hamiltonian systems. Namely, one replaces the classical algebra of observables, consisting of functions on a phase space, by a quantum Lie algebra of observables consisting of the same functions with the Poisson bracket instead of multiplication, and a unitary representation of this algebra. One can imagine the universal enveloping algebra of this Lie algebra (or its exponentiated form) as a noncommutative coordinate ring of the quantized initial phase space.

Into any concrete description of the quantum commutation rules enters a small parameter h (Planck's constant). (Of course, since it is not dimensionless, its "smallness" means rather that a characteristic action of a system of macroscopic scale is large when measured in units of h.) This means that one algebraic approach to noncommutative spaces and quantum groups is via deformation theory. The most intensively studied quantum groups up to now are represented by the deformations $U_h(\mathfrak{g})$ of the universal enveloping algebras of classical simple (or Kac–Moody) Lie algebras $\mathfrak{g}$.

We present an introduction to this approach to quantum groups in Section 3 of this chapter. Chapter 4 presents quantum groups as symmetry objects of quantum spaces.

Algebraically, there is a subtle difference between supergroups and quantum groups. Formally speaking, quantum groups are Hopf algebras, virtually noncommutative and noncocommutative, while supergroups are Hopf superalgebras, supercommutative if represented by the function rings. A difference in the axioms of these two types of objects is best understood if one writes them down in an abstract tensor category (cf. [DM]) and realizes that quantum groups are constructed over a tensor category of usual vector spaces, whereas supergroups are based upon a different tensor category of $\mathbb{Z}_2$-graded spaces with the twisted permutation isomorphism $S_{(12)} : V \otimes W \to W \otimes V$. One may well combine both variations and define quantum supergroups (cf. Chapter 4).

Section 4 of this chapter is devoted to some basic facts concerning monoidal, tensor, and pseudotensor categories whose role in understanding quantum groups stems also from a generalization of the Tannaka–Krein duality. Namely, a very important role is played by those quantum groups for which braid groups act naturally on tensor powers of representation spaces. Such an action is described by solutions of Yang–Baxter equations that emerged in two-dimensional statistical physics and recently became related also to two-dimensional conformal field theories.

1.3. FUNCTIONAL ANALYSIS. A functional-analytic approach, vigorously pursued in recent years by Alain Connes and his collaborators, starts with two remarks. First, due to Gelfand's theorem, cited earlier, one can take noncommutative C^*-algebras as a natural category for noncommutative topology. Second, there is a supply of quite common geometric situations leading to such algebras.

In [C2] and [C3], Connes suggests the following examples.

(a) Let V be a smooth manifold, F a smooth foliation on V. The leaf space, V/F, of course, exists as a topological space but is very far from being a manifold, and its properties cannot be described by conventional means. It is suggested that its topology is encoded in the C^*-algebra $C^*(V,F)$ defined, e.g., in [C4].

(b) Let Γ be a discrete group. The topology of the reduced dual space $\hat{\Gamma}$ is described by the norm closure of $\mathbb{C}\Gamma$ in the algebra of bounded operators in $l^2(\Gamma)$, that is, by a C^*-algebra $C_r^*(\Gamma)$.

(c) One can treat similarly the topology of quotient spaces V/Γ and V/G, where Γ (resp. G) is a discrete (resp. Lie) group acting upon a smooth manifold V. In these cases, the C^*-algebra in question is a crossed product of a function algebra of V with Γ (resp. G).

What kind of invariants of an algebra A should be qualified as topological invariants of an imaginary noncommutative space corresponding to A?

First, there are K-theoretical invariants. A general principle discovered first in algebraic geometry is that the topology of a usual ("commutative") space is encoded in the category of the vector bundles of this space, which in its turn is equivalent to a category of projective A-modules, where A is an appropriate function ring. K-groups are constructed directly in terms of this category.

Second, there are differential-geometric invariants embodied in the de Rham complex of A. This complex calculates (co)homology and contains some information on homotopical properties. K-theory and (co)homology are connected by the Chern character.

In [C3], Connes develops these ideas in a very broad context and, in particular, investigates the so-called cyclic cohomology groups of a ring, which he introduced in connection with his noncommutative de Rham complex. This very important construction was independently invented by B. L. Tsygan as an additive analog of K-theory (cf. [T], [FT]).

One of the latest most important developments is due to M. Wodzicki. His theory of noncommutative residue was reported in [Kas]; cf. his original work [W1]–[W3].

In Section 2 of this chapter, we review more algebraic parts of Connes's work, referring the reader to the bibliography for further reading.

1.4. ALGEBRAIC GEOMETRY. Turning to algebra-geometric sources of noncommutative geometry, one must confess that although its general influence was very significant, concrete endeavors to lay down foundations of noncommutative algebraic geometry Grothendieck-style were unsuccessful (but see [Ro]). One stumbling block invariably was noncommutative localization. The point is that whereas, say, a smooth manifold is described by its algebra of global smooth functions, an algebraic variety is not described by its algebra of polynomial functions unless it is affine. Hence, we must have functions that are defined only locally, and for this we probably need tangible geometric objects on which such functions are defined as local models. Notice that in Connes's approach, we have no local models: His C^*-algebras are connected with such spaces as V/F in a rather indirect way and are not readily visualized as functions on them.

Since attempts to glue together noncommutative algebraic spaces from affine ones generally fail, we have to resort to more particular cases and to learning lessons of other approaches.

For example, one can define some analogs of affine algebraic groups, following the lead of the theory of quantum groups, and study them as in classical algebraic geometry. In doing so, we discover that there are very special values of deformation parameters that, on the one hand, correspond to nontrivial representation theories and, on the other hand, lead to function rings that are almost commutative, that is, have large commutative subrings. One may try to use these subrings for constructing geometric spectra and localizing with respect to them so that the rest of the algebra becomes encoded in the structure sheaf on an actual space. Lessons of supergeometry may be quite helpful here.

Chapter 4 of this book will develop this viewpoint. For further reading, we recommend a very interesting recent work by B. Parshall and J.-P. Wang, "Quantum Linear Groups," Parts I and II, University of Virginia preprints, 1989.

2. Noncommutative de Rham Complex and Cyclic Cohomology

2.1. CYCLES. Following Connes [3], we shall call an *n-cycle* a triple $(\Omega, d, \int)$, where $\Omega = \bigoplus_{j=0}^{n} \Omega^j$ is a graded $\mathbb{C}$-algebra, d is its graded derivation of degree $1, d^2 = 0$, and $\int : \Omega^n \to \mathbb{C}$ is a closed graded (super)trace, i.e., a linear functional satisfying the following identities:

$$\int d\omega = 0 \quad \text{for all} \quad \omega \in \Omega^{n-1};$$

$$\int \omega\omega' = (-1)^{\deg(\omega)\deg(\omega')} \int \omega'\omega.$$

2.2. EXAMPLES. (a) Let X be a compact smooth oriented n-dimensional manifold, $(\Omega(X), d)$ its de Rham complex over $\mathbb{C}$, $\int$ the integral of volume forms over X. It is an n-cycle.

More generally, consider a closed q-current C on X. Then $(\bigoplus_{i=0}^{q} \Omega^i(X), d, \int = \langle C, \,.\, \rangle)$ is a q-cycle.

(b) For an associative $\mathbb{C}$-algebra A and a linear functional $\mathrm{tr} : A \to \mathbb{C}$ with $\mathrm{tr}([A, A]) = 0$, the triple $(A = \Omega = \Omega^0, \mathrm{tr})$ is a 0-cycle.

(c) By replacing $\int$ by $-\int$ in a cycle, we, by definition, change its *orientation*.

Direct sum of two cycles is defined in an obvious way. The following example describes a functional-analytic situation leading to cycles.

2.3. FREDHOLM MODULES. Let A be an associative $\mathbb{C}$-algebra, $H = H_0 \oplus H_1$ a $\mathbb{Z}_2$-graded separable Hilbert space, endowed with an odd bounded $\mathbb{C}$-linear involution F. Assume that H is also endowed with a structure of the left A-module such that A acts by even bounded operators.

Then (A, H, F) is called an *n-summable Fredholm A-module* if $[F, a] = Fa - aF \in L^n(H)$ for all $a \in A$, where $L^n(H)$ is the so-called *n-th Schatten ideal*, consisting of those bounded operators T, for which $|T|^n$ is of trace class, $|T| = (T^*T)^{1/2}$.

Given such a module, we can construct the following n-cycle. Put $\Omega^0 = A$; Ω^q = closure of the linear span in $L^{n/q}(H)$ of the family of operators

$(a^0 + \lambda \cdot \mathrm{id})[F,a^1][F,a^2]\dots[F,a^q]$ where $a^i \in A$, $\lambda \in \mathbb{C}$. Define d by $d\omega = i[F,\omega]$. Finally, for $\omega \in \Omega^n$, put

$$\int \omega = (-1)^n \operatorname{trace}(\omega).$$

Axiomatizing essential algebraic features of this situation, Connes arrives at the following abstract construction.

2.4. THE NONCOMMUTATIVE DE RHAM COMPLEX. For a fixed associative $\mathbb{C}$-algebra A, with or without identity, consider all algebra homomorphisms $f : A \to \Omega^0$, where Ω^0 is the 0-component of a differential graded unitary algebra (Ω, d). We do not assume that f transforms the identity of A (if any) into the identity of Ω^0.

These homomorphisms form a category with an initial object. Here is its direct construction.

$\Omega^0 = \hat{A} = A \oplus \mathbb{C}1$ (formal adjunction of identity);
$\Omega^n = \hat{A} \otimes A^{\otimes n}$ (tensor products over $\mathbb{C}$);
$d((a^0 + \lambda.1) \otimes a^1 \otimes \dots \otimes a^n) = 1 \otimes a^0 \otimes a^1 \otimes \dots \otimes a^n$;
in particular, $da = 1 \otimes a$ for $a \in A$;
the multiplication map $\Omega^m \otimes \Omega^n \to \Omega^{m+n}$ is uniquely defined by the following conditions: for $m = 0$, it is the standard $\hat{A}$- multiplication on the leftmost tensor factor of Ω^n, the Leibniz formula holds.

As an example, let us multiply $\hat{a}^0 \otimes a^1$ by $\hat{b}^0 \otimes b^1$, where $\hat{a}^0 = a^0 + \lambda \cdot 1$, $\hat{b}^0 = b^0 + \mu \cdot 1$:

$$\begin{aligned}(\hat{a}^0 \otimes a^1)(\hat{b}^0 \otimes b^1) &= (\hat{a}^0 da^1)(\hat{b}^0 db^1) = \hat{a}^0(da^1\hat{b}^0)db^1\\ &= \hat{a}^0[d(a^1\hat{b}^0) - a^1 d\hat{b}^0]db^1\\ &= \hat{a}^0 d(a^1\hat{b}^0)db^1 - (\hat{a}^0 a^1)db^0 db^1\\ &= \hat{a}^0 \otimes a^1 b^0 \otimes b^1 - \hat{a}^0 a^1 \otimes b^0 \otimes b^1.\end{aligned}$$

Here is a general formula for right multiplication by A:

$$(\hat{a}^0 \otimes a^1 \otimes \dots \otimes a^n)b = \sum_{j=0}^{n} (-1)^{n-j} \hat{a}^0 \otimes \dots \otimes a^j a^{j+1} \otimes \dots \otimes b.$$

Although this noncommutative de Rham complex has many properties in common with the usual one, it should not be considered as a "final solution." In fact, in Chapter 4, we shall see that in the category of quadratic algebras, for example, a natural substitute for the de Rham complex is one of the four Koszul complexes, having a very different structure.

2.5. TRACES. In order to complete a (truncated) de Rham complex $(\Omega^{\leq q}(A), d)$ to a cycle, we need a trace functional. If such a functional $\int : \Omega^q(A) \to \mathbb{C}$ is given, consider a linear map $\tau : A^{\otimes q} \to A^*$ (here the asterisk means linear dualization),

$$\tau(a^1 \otimes \ldots \otimes a^q)(a^0) = \int a^0 da^1 \ldots da^q.$$

Connes has shown that τ satisfies certain functional equations that can best be expressed by saying that τ is a cocycle of a very important complex, and that this correspondence between traces and cocycles is one-to-one.

To explain this result in a natural generality, we shall digress and start with the notion of a cyclic object of an abstract category.

2.6. CYCLIC OBJECTS. First recall that a simplicial object of a category C is a functor $\Delta^0 \to C$ where Δ is the category of well-ordered finite sets $[n] = \{0, 1, \ldots, n\}$ with nondecreasing maps as morphisms.

Following Connes, we similarly define a *cyclic object* of C as a functor $\Lambda^0 \to C$, where objects of Λ are

$$\{n\} = \text{ roots of unity of any degree } n + 1,$$

and morphisms are defined by any of the following equivalent ways.

By writing k instead of $\exp(2\pi i k/(n+1))$, introduce on {n} the cyclic order $0 \leq 1 \leq 2 \leq \cdots \leq n \leq 0$.

VARIANT 1. $\mathrm{Hom}_\Lambda(\{n\}, \{m\})$ = the set of homotopy classes of continuous cyclically nondecreasing maps $\varphi : S^1 \to S^1$ such that $\varphi(\{n\}) \subseteq \{m\}$. Here $S^1 = \{z \in \mathbb{C} \mid |z| = 1\}$, and homotopy is considered in the same class of maps.

VARIANT 2. A morphism $\{n\} \to \{m\}$ is a pair consisting of a set-theoretical map $f : \{n\} \to \{m\}$ and a set σ of total orders, one on each fiber $f^{-1}(i), i \in \{m\}$. They must satisfy the following condition: The cyclic order on $\{n\}$ induced by the standard cyclic order on $\{m\}$, and σ should coincide with the standard cyclic order on $\{n\}$.

The composition rule is: $(g, \tau)(f, \sigma) = (gf, \tau\sigma)$, where $i \leq j$ with respect to $\tau\sigma$ if either $f(i) \leq f(j)$ with respect to τ, or $f(i) = f(j)$ and $i < j$ with respect to σ.

A given f can be extended to a morphism in Λ iff it is cyclically nondecreasing. This extension is unique unless f is constant, in which case there are $n + 1$ extensions.

VARIANT 3. Morphisms in Λ are given formally by the following sets of generators and relations.

Generators:

$$\delta_n^i : \{n-1\} \to \{n\}; \quad \sigma_n^i : \{n+1\} \to \{n\}; \quad \tau_n : \{n\} \to \{n\}.$$

Relations:

$$\delta_n^j \delta_{n-1}^i = \delta_n^i \delta_{n-1}^{j-1} \quad \text{for} \quad i \leq j;$$

$$\sigma_n^j \sigma_{n+1}^i = \sigma_n^i \sigma_{n+1}^{j+1} \quad \text{for} \quad i \leq j;$$

$$\sigma_n^j \delta_{n+1}^i = \begin{cases} \delta_n^i \sigma_{n-1}^{j-1} & \text{for} \quad i < j; \\ \text{id} & \text{for} \quad i = j \text{ or } j+1; \\ \delta_n^{i-1} \sigma_{n-1}^j & \text{for} \quad i > j+1; \end{cases}$$

$$\tau_n \delta_n^i = \delta_n^{i-1} \tau_{n-1} \quad \text{for} \quad i = 1, \ldots, n;$$

$$\tau_n \sigma_n^i = \sigma_n^{i-1} \tau_{n+1} \quad \text{for} \quad i = 1, \ldots, n;$$

$$\tau_n^{n+1} = \text{id}.$$

In terms of the previous description, ∂_n^i omits i; σ_n^i takes the value i twice; $\tau_n(j) = j+1$. Only for σ_0^0, we must fix an order on a fiber: it is $0 < 1$.

Remarkably, Λ is isomorphic to Λ^0. The isomorphism is identical on objects and acts as follows on morphisms: If, in the second description, $(f, \sigma) : \{n\} \to \{m\}$, we define $(f, \sigma)^* = (g, \tau) : \{m\} \to \{n\}$ by

$g(i) = \sigma$ – minimal element of $f^{-1}(j)$ where j is the maximal element of $f^{-1}(j)$ cyclically preceding i.

An additional piece of information is needed only if g is constant. This happens precisely when f is constant. Then τ is defined by the condition that f is the τ-minimal element of $\{n\}$.

2.7. CYCLIC COMPLEXES. Let now $\mathbf{E} = (E_n, d_i^n, s_i^n, t_n)$ be a cyclic object of an *abelian category*, where d, s, t, respectively, correspond to δ, σ, τ. Put

$$d^n = \sum_{i=0}^{n} (-1)^i d_i^n : E_n \to E_{n-1};$$

$$d'^n = \sum_{i=0}^{n-1} (-1)^i d_i^n : E_n \to E_{n-1};$$

$$t = (-1)^n t_n : E_n \to E_n; N = \sum_{i=0}^{n} t^i.$$

First of all, (E_n, d^n) and (E_n, d'^n) are complexes. Since $d(1-t) = (1-t)d'$, they can be combined in the following bicomplex:

$$\begin{array}{ccccccc}
\vdots & & \vdots & & \vdots & & \\
d\downarrow & & -d'\downarrow & & d\downarrow & & \\
E_2 & \xleftarrow{1-t} & E_2 & \xleftarrow{N} & E_2 & \xleftarrow{1-t} & \dots \\
d\downarrow & & -d'\downarrow & & d\downarrow & & \\
E_1 & \xleftarrow{1-t} & E_1 & \xleftarrow{N} & E_1 & \xleftarrow{1-t} & \dots \\
d\downarrow & & -d'\downarrow & & d\downarrow & & \\
E_0 & \xleftarrow{1-t} & E_0 & \xleftarrow{N} & E_0 & \xleftarrow{1-t} & \dots
\end{array}$$

Denote by $C \cdot E$ the associated complex and define the *cyclic homology* $HC \cdot (\mathbf{E})$ of the cyclic object $\mathbf{E}$ as $H(C \cdot E)$.

If multiplication by $n+1$ is an isomorphism of E_n for all n, the rows of the bicomplex are exact everywhere except the leftmost column. Thus, in this case, $HC \cdot (\mathbf{E}) = H(E \cdot /(1-t)\mathbf{E}\cdot)$.

A *cocyclic object* of C can be defined dually as a functor $\mathbf{E} : \Lambda \to C$. If C is abelian, one can repeat the previous construction with the arrows reversed. In this way, we obtain the *cyclic cohomology* $HC^{\cdot}(\mathbf{E})$.

2.8. Cyclic Cohomology as a Derived Functor. Suppose that C is the category of k-modules over a commutative ring k. Denote by Λk an object of the category ΛC of cocyclic objects of C, all of whose components are k and all of whose morphisms are identical.

ΛC is an abelian category, and

$$HC^k(\mathbf{E}) = \mathrm{Ext}^k_{\Lambda C}(\Lambda k, \mathbf{E}).$$

One can similarly treat cyclic homology as a certain Tor$\cdot$-functor. These results can be generalized to the abstract categories admitting a unit object 1 (similar to k) and internal Hom of diagrams.

2.9. Connection with the Hochschild Homology. Let A be an algebra over a commutative ring containing $\mathbb{Q}$. We can define a cyclic object $\mathbf{A}$ with the ith component $A^{\otimes(i+1)}$ and the following structure morphisms:

$$d_i^n(a^0 \otimes \dots \otimes a^n) = a^0 \otimes \dots \otimes a^i a^{i+1} \otimes \dots \otimes a^n, \quad 0 \le i \le n;$$
$$d_n^n(a^0 \otimes \dots \otimes a^n) = a^n a^0 \otimes a^1 \otimes \dots \otimes a^{n-1};$$
$$s_i^n(a^0 \otimes \dots \otimes a^n) = a^0 \otimes \dots \otimes a^i \otimes 1 \otimes a^{i+1} \otimes \dots \otimes a^n;$$
$$t_n(a^0 \otimes \dots \otimes a^n) = a^n \otimes a^0 \dots \otimes a^{n-1}.$$

The left column of the bicomplex associated to this object in Section 2.7 is called the Hochschild complex $C \cdot (A, A)$.

Let us denote by L the whole bicomplex, by S its endomorphism shifting it by two columns to the right, by K the bicomplex consisting of the first two columns of L and zeroes elsewhere. There is an exact sequence

$$0 \to K \to L \to L/K \to 0$$

and canonical isomorphisms $H \cdot (\Delta K) = H \cdot (A, A), S : L \simeq L/K$. Passing to homology, we get an exact sequence:

$$\ldots \to H_n(A, A) \to HC_n(A) \xrightarrow{S} HC_{n-2}(A) \xrightarrow{\delta} H_{n-1}(A, A) \to \ldots .$$

Here we write $HC_n(A)$ instead of $HC_n(\mathbf{A})$.

For cyclic cohomology, one can obtain in the same way an exact sequence involving Hochschild cohomology with coefficients in A^*:

$$\ldots \to H^n(A, A^*) \to HC^{n-1}(A) \to HC^{n+1}(A) \to h^{n+1}(A, A^*) \to \ldots .$$

2.10. Relative Cyclic Homology of Algebras. Consider a differential graded algebra A over a field k of characteristic zero. It is called *free* if it is isomorphic to the tensor algebra of a graded vector space (nothing is assumed about the differential). More generally, a morphism $A \to B$ is called free if it is isomorphic to a morphism $A \to A * C$, where C is free and $*$ denotes amalgamation. We also say that B is free over A.

The category of associative k-algebras is embedded in the category of differential graded algebras with vanishing components of degree zero.

Let $f : A \to B$ be a morphism of k-algebras. A *resolution* of B over A is a commutative diagram

$$\begin{array}{ccc} & & R \\ & {}^{i}\nearrow & \\ A & & \downarrow \pi \\ & \searrow^{f} & \\ & & B \end{array}$$

such that R is free over A and π is a surjective quasiisomorphism. Every f admits a resolution.

Consider a resolution R as a complex. From the Leibniz formula, it follows that $[R, R] + i(A)$ is a subcomplex (here $[R, R]$ is the linear span of supercommutators). Put

$$HC_n(A \to B) = H_{n+1}(R/([R, R] + i(A))).$$

We then have the following results.

(a) These homology groups do not depend on the choice of a resolution and define a covariant functor on the category of morphisms of k-algebras.

(b) Composition of morphisms $A \to B \to C$ generates a functorial exact sequence

$$\begin{aligned} \ldots &\to HC_n(A \to B) \to HC_n(A \to C) \\ &\to HC_n(B \to C) \to HC_n(A \to B) \to \ldots \end{aligned}$$

(c) There are functorial isomorphisms $HC_n(A \to 0) = HC_n(A)$.

2.11. CONNECTION WITH THE DE RHAM COHOMOLOGY. Let A be a finitely generated commutative k-algebra, and k a field of characteristic zero. One can define the algebraic de Rham complex $(\Omega A, d)$ in the standard way.

When $k = \mathbb{C}$ and A is the ring of polynomial functions on a nonsingular affine variety X, one can identify $H^{\cdot}(\Omega A)$ with the singular cohomology $H^{\cdot}(X, \mathbb{C})$. In the general case, to calculate $H^{\cdot}(X, \mathbb{C}), X = \mathrm{Spec}(A)(\mathbb{C})$ algebraically, one must first choose an affine embedding of $\mathrm{Spec}(A)$, i.e., a surjection $B \to A$, where B is a polynomial algebra. Let I be the kernel of this surjection. Define the following filtration of the complex ΩB:

$$F^n \Omega^j B = \begin{cases} \Omega^j B & \text{for } n \leq j; \\ I^{n-j} \Omega^j B & \text{for } n > j. \end{cases}$$

Put

$$H^{\cdot}_{\mathrm{cris}}(A; n) = H^{\cdot}(\Omega B / F^{n+1} \Omega B).$$

These crystalline cohomology groups do not depend on the choice of $B \to A$. Grothendieck established an isomorphism

$$H^{\cdot}(X, \mathbb{C}) = H^{\cdot}(\varprojlim \Omega B / F^{n+1} \Omega B).$$

Cyclic homology is related with the finite levels of this filtration via functorial morphisms,

$$\chi_{n,i} : HC_n(A) \to H^{n-2i}_{\mathrm{cris}}(A, n-1).$$

If I is generated by a regular sequence, we have the isomorphisms

$$\bigoplus_i \chi_{n,i} : HC_n(A) \to \bigoplus_{0 \leq 2i \leq n} H^{n-2i}_{\mathrm{cris}}(A; n-i).$$

If $\mathrm{Spec}(A)$ is a reduced smooth scheme, we have

$$H^n_{\mathrm{cris}}(A; m) = \begin{cases} H^n(\Omega A) = H^n_{\mathrm{DR}}(\mathrm{Spec}(A)) & \text{for } n \leq m; \\ \Omega^n A / d\Omega^{n-1} A & \text{for } n = m. \end{cases}$$

Thus, in this case cyclic homology is

$$HC_n(A) \simeq \Omega^n A / d\Omega^{n-1} A \oplus \left(\bigoplus_{i \geq 1} H_{\mathrm{DR}}^{n-2i}(\mathrm{Spec}(A)) \right).$$

2.12. TRACES AND CYCLIC COCYCLES. Let us now return to the situation in Section 2.5, where we started with a linear functional $\int : \Omega^{\leq q}(A) \to \mathbb{C}$, completing the complex $(\Omega^{\leq q}, d)$ to a Connes cycle and constructed an operator $\tau : A^{\otimes q} \to A^*$. Here $\Omega(A)$ is the noncommutative de Rham complex.

One can now explain in what sense cyclic cohomology classifies integrals: *the correspondence* $\int \leftrightarrow \tau$ *is a bijection between closed graded traces and cyclic cocycles (called characters).*

We can now extend some further notions of topology and differential geometry to the noncommutative case.

2.13. COBORDISM. An $(n+1)$-*chain* in Connes's sense is a quadruple $(\Omega, \partial\Omega, r, \int)$ consisting of the following data.

(a) $\Omega = \bigoplus_{i=0}^{n+1} \Omega^i$, $\partial\Omega = \bigoplus_{i=0}^{n} (\partial\Omega)^i$ are differential graded algebras, and $r : \Omega \to \partial\Omega$ is a surjective morphism of degree zero.

(b) $\int : \Omega^{n+1} \to \mathbb{C}$ is a trace such that $\int d\omega = 0$ if $r(\omega) = 0$.

Given a chain, its *boundary* is the n-cycle $(\partial\Omega, d, \int')$, where $\int' \omega' = \int d\omega$ if $r(\omega) = \omega'$.

Two cycles Ω', Ω'' are called cobordant if there exists a chain with boundary $\Omega' \oplus \bar{\Omega}''$, where the bar denotes the orientation reversal.

In the same way, one can define the relative notions of cobordism over an algebra A.

Cobordism is an equivalence relation.

It can be determined in terms of cyclic and Hochschild cohomology. Let τ', τ'' be the characters of cycles Ω', Ω''. Then these cycles are cobordant iff the difference of their cohomology classes belongs to the image of the morphism described in Section 2.9.

$$S : H^{n+1}(A, A^*) \to HC^n(A).$$

2.14. CONNECTIONS. Consider a cycle $\rho : A \to \Omega$ over A and a projective right A-module E of finite rank. A *connection* on E is a $\mathbb{C}$-linear map $\nabla : E \to E \otimes_A \Omega^1$ with the following property:

$$\nabla(ea) = (\nabla e)a + a \otimes d(\rho(a)) \quad \text{for all } e \in E, a \in A.$$

We shall assume that A has an identity and put $\mathcal{E} = E \otimes_A \Omega$. We extend ∇ to $\mathcal{E}$ by

$$\nabla(e \otimes \omega) = (\nabla e)\omega + e \otimes d\omega.$$

Consider the graded $\mathrm{End}_A(E)$-algebra $\mathrm{End}_\Omega(\mathcal{E})$. For $T \in \mathrm{End}(\mathcal{E})$ put

$$\delta(T) = \nabla T - (-1)^{\deg(T)} T \nabla.$$

Define a trace functional on $\mathrm{End}_\Omega(\mathcal{E})^n$ as the composition of the matrix trace and $\int : \Omega^n \to \mathbf{C}$.

The triple $(\mathrm{End}_\Omega(\mathcal{E}), \delta, \int \circ \mathrm{tr})$ is not, however, a cycle because $\delta^2 \neq 0$ in general. In fact, ∇^2 on $\mathcal{E}$ is the multiplication by a curvature form θ, and $\delta^2 T = [\theta, T]$.

In noncommutative geometry, there is a universal method of killing curvatures.

Abstractly, consider a system $(\Xi, \delta, \theta, \int)$ consisting of a graded differential algebra (Ξ, δ) with the last component Ξ^n, a closed trace $\int$, and an element $\theta \in \Xi^2$ such that $\delta\theta = 0$ and $\delta T = [\theta, T]$ for all $T \in \Xi$.

We adjoin to Ξ an element X of degree 1 subject to the following relations:

$$X^2 = \theta; \quad \omega_1 X \omega_2 = 0 \quad \text{for all} \quad \omega_i \in \Xi.$$

On the resulting algebra Ξ' define d' and $\int'$ by the following rules:

$$d'\omega = \delta\omega + X\omega - (-1)^{\deg(\omega)}\omega X \quad \text{for} \quad \omega \in \Xi;$$
$$d'X = 0;$$

$$\int' (\omega_{11} + \omega_{12}X + X\omega_{21} + X\omega_{22}X) = \int \omega_{11} - (-1)^{\deg(\omega_{11})} \int \omega_{22}\theta,$$

where $\deg(\omega_{11}) = n = \deg(\omega_{12}) + 1 = \deg(\omega_{22}) + 2$.

Connes proves that $(\Xi', d', \int')$ is a cycle.

If we now apply this construction to $(\mathrm{End}_\Omega(\mathcal{E}), \delta, \theta, \int)$, we get a cycle over $\mathrm{End}_A(E)$. Its character depends only on the class of E in $K_0(A)$ and on the character of the initial cycle. Besides, in the spirit of Morita, the cyclic cohomology $H(A)$ of A and $\mathrm{End}_A(E)$ coincide.

Connes proves that the resulting map $K_0(A) \times H(A) \to H(A)$ is biadditive and makes of $H(A)$ a $K_0(A)$-module if A is commutative.

3. Quantum Groups and Yang–Baxter Equations

3.1. AFFINE ALGEBRAIC GROUPS. An affine algebraic group G over a field k can be defined in the following "naive" way. It is given by an ideal $J \subset k[z_i^j], i, j = 1, \ldots, n$ with the following properties.

(a) Let $U = (u_i^j), V = (v_i^j)$ be two $n \times n$-matrices whose coefficients lie in a commutative k-algebra A and satisfy the algebraic equations $f(u_i^j) = f(v_i^j) = 0$ for all $f \in I$ (or for a family of generators of I). Then UV is also a solution of this system.

(b) The identity matrix $I_n = I = (\delta_i^j)$ satisfies J.

(c) If U (as in (a)) satisfies J and is invertible in $M(n, A)$, then U^{-1} also satisfies J.

In more invariant terms, G is determined by its functor of points on the category of commutative k-algebras A,

$$G(A) = \{f \in \mathrm{Hom}(k[z_i^j]/J, A) \mid (f(z_i^j)) \text{ is invertible}\}.$$

Of course, the invertibility condition can be built into the definition of the function ring of G: It suffices to invert formally $Z = (z_i^j)$ first, that is, to introduce new independent variables (y_i^j) forming an $n \times n$ matrix Y, and to put

$$F(G) = k[z_i^j, y_i^j]/(J, \text{ coefficients of } XY - I).$$

Then

$$G(A) = \mathrm{Hom}_{k-\mathrm{alg}}(F(G), A).$$

The initial ring $F(\overline{G}) = k[z_i^j]/J$ represents an *algebraic matrix semigroup* $\overline{G}$. The matrix Z itself, representing the identity morphism of $F(G)$ or $F(\overline{G})$, is called a *generic point of* G or $\overline{G}$. An arbitrary point corresponding to an *injective* morphism $F(G) \to A$(resp. $F(\overline{G}) \to A$) is also called generic.

Looking at conditions (a), (b), and (c) in terms of generic points, we can rewrite them in the following way. For U, V take points $F(G) \to F(G) \otimes F(G)$ corresponding to $z_i^j \to z_i^j \otimes 1$ and $z_i^j \to 1 \otimes z_i^j$, respectively. Then the product corresponds to the point $Z \otimes Z$ with coefficients $(Z \otimes Z)_i^k = \sum z_i^j \otimes z_j^k$ (thus, our tensor product of matrices is not Kronecker's tensor product!). In other words, we get the diagonal map (or comultiplication) $\Delta : F(G) \to F(G) \otimes F(G), \Delta(Z) = Z \otimes Z$. Similarly, condition (b) furnishes a counit map $\varepsilon : F(G) \to k$, and condition (c) furnishes an antipode $i : F(G) \to F(G) : i(Z) = Y = Z^{-1}$.

It is well known that if we explicitly add to this data the multiplication map $m : F(G) \otimes F(G) \to F(G)$ and the unit map $\eta : k \to F(G)$, we shall obtain a particular case of a general notion of Hopf algebra. Let us recall its definition.

3.2. Algebras, Coalgebras, Bialgebras, and Hopf Algebras.

(a) *An associative k-algebra with unit* is a linear space E with the structure maps $m : E \otimes E \to E$ and $\eta : k \to E$ such that

$$m \circ (m \otimes \mathrm{id}) = m \circ (\mathrm{id} \otimes m) : E \otimes E \otimes E \to E;$$
$$m \circ (\eta \otimes \mathrm{id}) = m \circ (\mathrm{id} \otimes \eta) = \mathrm{id} : k \otimes E = E \otimes k = E \to E.$$

(b) *A coassociative k-algebra with counit* is a linear space E with structure maps $\Delta : E \to E \otimes E$ and $\varepsilon : E \to k$ such that

$$(\mathrm{id} \otimes \Delta) \circ \Delta = (\Delta \otimes \mathrm{id}) \circ \Delta : E \to E \otimes E \otimes E;$$
$$(\varepsilon \otimes \mathrm{id}) \circ \Delta = (\mathrm{id} \otimes \varepsilon) \circ \Delta = \mathrm{id} : E \to E = k \otimes E = E \otimes k.$$

(c) *A bialgebra* is a linear space E endowed with the structures of an algebra (m, ε) and a coalgebra (Δ, η) satisfying a compatibility axiom that can be written in the form: *Δ is a morphism of k-algebras.*

It is assumed here that the multiplication in $E \otimes E$ is given by the usual rule $(e \otimes f)(e' \otimes f') = ee' \otimes ff'$.

This is the main place where the definition of, say, superbialgebra differs from that of bialgebra: A sign enters in the formula for multiplication in $E \otimes E$. For this reason, it is better to write the compatibility axiom in the form

$$(m \otimes m) \circ S_{(23)} \circ (\Delta \otimes \Delta) = \Delta \circ m : E \otimes E \to E \otimes E,$$

where $S_{(23)} : E^{\otimes 4} \to E^{\otimes 4}$ is the morphism interchanging two middle factors. It may become nontrivial in a tensor category different from that of vector spaces, e.g., that of $\mathbb{Z}_2$-graded vector spaces. If we carefully write all the relevant axioms with the necessary permutation morphisms, they will be automatically applicable in more general tensor categories (cf. Section 3.3).

We see also that the bialgebra data and axioms are self-dual with respect to the reversal of all arrows and the replacement of (m, η) by (Δ, ε), and vice versa.

(d) *A Hopf algebra* is a bialgebra $(E, m, \eta, \Delta, \varepsilon)$ endowed with a linear map $i : E \to E$ (*antipode*) such that

$$m \circ (i \otimes \mathrm{id}) \circ \Delta = m \circ (\mathrm{id} \otimes i) \circ \Delta = \eta \circ \varepsilon : E \to E.$$

The properties of an antipode in a general Hopf algebra can differ considerably from those in an affine algebraic group.

First, it is in general not a morphism of algebras or coalgebras; it reverses both multiplication and comultiplication. Precisely, put $m^{\mathrm{op}} = m \circ S_{(12)}, \Delta^{\mathrm{op}} = S_{(12)} \circ \Delta$. Reversing in a bialgebra either multiplication, comultiplication, or both, we still get a bialgebra. An antipode i, if it exists at all, is a bialgebra morphism $(E, m, \eta, \Delta, \varepsilon) \to (E, m^{\mathrm{op}}, \eta, \Delta^{\mathrm{op}}, \varepsilon)$. If, in addition, it is bijective (which is not always so), then i^{-1} is an antipode for $(E, m^{\mathrm{op}}, \Delta)$ and $(E, m, \Delta^{\mathrm{op}})$, hence i is one for $(E, m^{\mathrm{op}}, \Delta^{\mathrm{op}})$.

E is commutative if $m = m^{\mathrm{op}}$; cocommutative if $\Delta = \Delta^{\mathrm{op}}$.

If an antipode for a bialgebra exists, it is unique (cf. [A] and Chapter 4), but not necessarily bijective. If it is bijective, it may have arbitrary finite or infinite order.

One can now easily prove that an affine algebraic group is (the spectrum of) a finitely generated (as algebra) commutative Hopf algebra, and vice versa. The group itself is commutative iff this Hopf algebra is cocommutative.

Now a tentative definition of an affine group in noncommutative geometry, or quantum group, is obvious: It should be a Hopf algebra, in general noncommutative and noncocommutative, with some finiteness conditions and possibly a condition of bijectivity of the antipode.

One remark is in order now. We have seen that the reversal of arrows transforms a bialgebra into a bialgebra and an antipode into an antipode. This formal duality can be transformed into a linear duality functor $(E, m, \Delta) \to (E', \Delta', m')$, where E' consists of linear functionals on E vanishing on an ideal of finite codimension. For an affine algebraic group, $(F(G))'$ consists of distributions with finite support. In characteristic zero, the distributions supported by identity form the universal enveloping algebra of the corresponding Lie algebra. In general, speaking of a quantum group, we must specify how we imagine a given Hopf algebra: as its function algebra or its distribution algebra. We shall usually prefer the first choice and state our definitions correspondingly. In particular, speaking of *representations*, we shall deal with *comodules* rather than modules (cf. below).

3.3. AFFINE QUANTUM GROUPS IN NONCOMMUTATIVE GEOMETRY. We shall often construct our groups by direct generalization of the data in Section 3.1, therefore we shall start by rephrasing them in the noncommutative situation. Consider a bilateral ideal $J \subset k\langle z_i^j \rangle, i, j = 1, \dots, n$, where $k\langle z_i^j \rangle$ is the free associative algebra. We shall say that J defines a *quantum matrix semigroup* $\overline{G}$, with the function ring $F(\overline{G}) = k\langle z_i^j \rangle / J$, if the following analogs of the conditions 3.1(a) and (b) are valid:

(a′) Let $U = (u_i^j), V = (v_i^j)$ be two $n \times n$ matrices whose coefficients lie in an associative k-algebra A, satisfy the noncommutative polynomial equations $f(u_i^j) = f(v_i^j) = 0$ for all $f \in J$ and furthermore pairwise commute: $[u_i^j, v_k^l] = 0$. Then UV also verifies $f((UV)_i^j) = 0$.

Note the commutation condition: Two points of a quantum (semi)-group can be multiplied only if their coefficients are "simultaneously pairwise observable." In particular, one cannot define a "one-parametric subgroup": one cannot even hope that U^2 or U^{-1} are points of the same group!

(b′) The identity matrix $I_n = I$ satisfies J.

Of course, this is equivalent to the statement that $(F(\overline{G}), \Delta, \varepsilon)$ is a bialgebra, where $\Delta(Z) = Z \otimes Z, \varepsilon(Z) = I$ (Δ and ε are applied coefficientwise).

For the reasons that should be already clear, we must replace 3.1(c) by a more complicated condition if we want to go from a semigroup to a group.

If (E, m, Δ) already admits an antipode, it represents a quantum group. Otherwise, one can argue as follows.

In the category of morphisms of bialgebras $F(\overline{G}) \to H$, where H is a Hopf algebra, there is a universal morphism. The corresponding Hopf algebra represents a quantum group G, and A-points of this group is $\mathrm{Hom}_{k-\mathrm{alg}}(H, A)$.

We shall prove this in Chapter 4; to construct H from $F(\overline{G})$, it is necessary formally to invert $Z, (Z^{-1})^t, (((Z^{-1})^t)^{-1})^t, \ldots$, etc., to infinity.

Therefore, an analog of Section 3.1(c). must ask for invertibility of an infinite set of matrices, and also for the validity of a set of noncommutative polynomial equations for their elements.

Since in our presentation a special role is played by a matrix Z, it is appropriate to clarify its place in the theory. For an affine algebraic group, the choice of Z is equivalent to that of a faithful representation of G in the coordinated vector space k^n. The same is true in general if one replaces representations by comodules.

3.4. CATEGORY OF COMODULES. A left (resp. right) comodule M can be defined over any coalgebra (E, Δ, ε). The data and the axioms can be obtained by reversing arrows from those of a module: M is a k-space endowed with comultiplication $\delta : M \to E \otimes M$ (resp. $M \to M \otimes E$) such that

$$(\Delta \otimes \mathrm{id}) \circ \delta = (\mathrm{id} \otimes \delta) \circ \delta : M \to E \otimes E \otimes M \quad \text{(coassociativity)};$$
$$(\varepsilon \otimes \mathrm{id}) \circ \delta = \mathrm{id} : M \to k \otimes M = M \quad \text{(counit)},$$

and similarly for the right coaction.

A morphism of left comodules is a linear map $r : M \to N$ such that $(\mathrm{id} \otimes r) \circ \delta_M = \delta_N \circ r$. Right comodules are defined in a similar manner. The direct sum is defined in a straightforward way. Left and right are exchanged by going to Δ^{op}: If (M, δ) is a left (right) (E, Δ)-module, $(M, \delta^{\mathrm{op}} = S_{(12)} \circ \delta)$ is a right (left) $(E, \Delta^{\mathrm{op}})$-module. This is an isomorphism of categories. Checking this involves only axioms of the basic tensor category, hence is valid also for supercoalgebras, etc.

In order to define the *tensor product* of two left (resp. right) comodules, we need multiplication. If (E, m, Δ) is a bialgebra, and M, N are two left comodules, we define the coaction map

$$\delta_{M \otimes N} : M \otimes N \to E \otimes M \otimes N$$

as the composition

$$M \otimes N \xrightarrow{\delta_M \otimes \delta_N} E \otimes M \otimes E \otimes N$$
$$\xrightarrow{S_{(23)}} E \otimes E \otimes M \otimes N \xrightarrow{m \otimes \mathrm{id} \otimes \mathrm{id}} E \otimes M \otimes N.$$

Again, this is well defined in view of the general tensor category axioms.

Finally, if M is a left (E, Δ)-comodule, the dual space M^* has a natural structure of the right comodule, which can be transformed into a left $(E, \Delta^{\mathrm{op}})$-comodule. If $(E, \Delta^{\mathrm{op}})$ is (a part of) a Hopf algebra, we can use the antipode $i : (E, \Delta^{\mathrm{op}}) \to (E, \Delta)$ to induce on M^* the structure of (E, Δ) comodule.

Thus, in the category of, say, left comodules over a Hopf algebra, one can define the data of a tensor category, but in general, they will not satisfy the usual axioms, e.g., commutativity of the tensor product. We shall see it in down-to-earth terms after discussing the connection with multiplicative matrices.

Let (E, Δ, ε) be a coalgebra. A matrix $Y \in M(n, E)$ is called *multiplicative*, if

(a) $\Delta(Y) = Y \otimes Y$.

(b) $\varepsilon(Y) = I_n$.

Let (k^n, δ) be a left E-comodule. Define $Y = (y_i^j) \in M(n, E)$ by $\delta(e_i) = \sum_j y_i^j \otimes e_j$, where $\{e_j\}$ is the standard basis of k^n.

3.5. PROPOSITION. *(a) The construction described above establishes a bijection between $n \times n$-multiplicative matrices and structures of left comodule on k^n.*

(b) Let $r : (k^m, \delta_1) \to (k^n, \delta_2)$ be a morphism of E-comodules, given by a matrix $R = (r_i^j) : r(e_i) = \sum r_i^j \otimes e_j'$. Then $Y_1 R = R Y_2$, where Y_i are the multiplicative matrices defining the comodules. Any $R \in M(m \times n, k)$ with this property represents such a morphism.

(c) The direct sum of comodules represented by Y_1, Y_2 is represented by $\begin{pmatrix} Y_1 & 0 \\ 0 & Y_2 \end{pmatrix}$.

(d) The tensor product of comodules represented by Y_1, Y_2 is represented by the Kronecker product $Y_1 \odot Y_2$;

$$(Y_1 \odot Y_2)_{ij}^{kl} = (Y_1)_i^k (Y_2)_j^l.$$

(e) Y is multiplicative for $(E, \Delta, \varepsilon) \Leftrightarrow Y^t$ is multiplicative for $(E, \Delta^{\mathrm{op}}, \varepsilon)$.

All these statements directly follow from the definitions. Note only that statement (d) changes in the category of superspaces: Some signs must enter in the definition of the Kronecker product (see Chapter 4).

One can say that multiplicative matrices form a category equivalent to that of finite-dimensional left comodules, with morphisms defined by (b).

Consider now the following situation. Let $M_i, i = 1, 2.3$ be three left comodules over a bialgebra. Then the canonical isomorphism of the linear spaces $(M_1 \otimes M_2) \otimes M_3 \to M_1 \otimes (M_2 \otimes M_3)$ is also an isomorphism of comodules. This follows from the coassociativity axiom of Δ. However, the canonical isomorphism $M_1 \otimes M_2 \to M_2 \otimes M_1$ in general is not an isomorphism of comodules. In fact, it may well happen that these comodules are not isomorphic at all. For example, one-dimensional comodules correspond to multiplicative (or grouplike) elements that may well form a noncommutative group as in the Hopf algebra of functions on a finite non-abelian group.

Suppose, however, that for a comodule M, there exists a nontrivial isomorphism (or just a morphism) $r : M^{\otimes 2} \to M^{\otimes 2}$. If M is given by a matrix Z and r is given by R, we get from 3.5(b) and (d) a quadratic relation between the coefficients of Z,

$$(3.1) \qquad RZ \odot Z = Z \odot ZR.$$

3.6. Yang–Baxter Equations and Braid Groups. The standard isomorphism $S_{(12)} : M \otimes M \to M \otimes M$ can be used to define a representation of the symmetric group S_n on $M^{\otimes n}$: One decomposes a permutation into a product of transpositions and then takes the product of the corresponding linear operators. The result does not depend on the choice of the decomposition because the Coxeter relations

$$S_{(12)} S_{(23)} S_{(12)} = S_{(23)} S_{(12)} S_{(23)},$$
$$S_{(12)}^2 = \mathrm{id}$$

form a presentation of S_3. The similar relations for an arbitrary linear operator $R \in \mathrm{End}(M \otimes M)$ (where $R_{12} = R \otimes \mathrm{id}$, etc.)

$$(3.2) \qquad R_{12} R_{23} R_{12} = R_{23} R_{12} R_{23} : M^{\otimes 3} \to M^{\otimes 3}$$

are called the (quantum) Yang–Baxter equation. An invertible solution of this equation allows us to define an action of the Artin braid group Bd_n on $M^{\otimes n}$. If in addition

$$(3.3) \qquad R_{12}^2 = \mathrm{id}$$

("unitarity condition"), this action reduces to that of S_n.

L. D. Faddeev and his collaborators use the relations (3.2) and (3.1) as a basic definition for the class of quantum groups they consider (see [ReTF]). Starting from a Yang–Baxter operator R, they construct a bialgebra

$$E = k\langle z_i^j \rangle / (\text{coefficients of } RZ \odot Z - Z \odot ZR).$$

(The last stage, transition from E to its Hopf envelope, is discussed in [ReTF] in a less general setting, where the noncommutative localization can be replaced by a commutative one.)

It is worth mentioning that the coefficients of $RZ \odot Z - Z \odot ZR$ generate a coideal with respect to $\Delta(Z) = Z \otimes Z$ for any R, so that in the construction of E, the Yang–Baxter property plays no role at all.

3.7. SOME VARIATIONS. Since the Yang–Baxter operators play the role of the structure constants of quantum (semi)groups with nice tensorial properties of the representation category, it is natural to discuss here various approaches to their classification.

We shall briefly comment upon some directions of recent research.

3.7.a. Classical Yang–Baxter Equations. Consider a Yang–Baxter operator R close to $S_{(12)}$, say, $R = S_{(12)} + hS_{(12)}r + 0(h^2)$, where h is a small parameter. By inserting this into Eq. (3.2) and considering the equation modulo h^3, we get the following *classical YB-equations for a linear operator* $r \in \mathrm{End}(E \otimes E) = \mathrm{End}(E)^{\otimes 2}$:

(3.4) $$[r_{12}, r_{13}] + [r_{12}, r_{23}] + [r_{13}, r_{23}] = 0.$$

Here, say, $r_{12} = r \otimes \mathrm{id}$, and the commutator of r_{12} and r_{13} refers to their common first factor in $\mathrm{End}(E)^{\otimes 3}$.

V. G. Drinfeld considered Eq. (3.4) as an abstract equation in a Lie algebra $\mathfrak{g}$, related it to various beautiful structures in the classical Lie group theory (Poisson–Lie groups; isotropic triples), and discussed the quantization of a classical solution, i.e., its extension modulo growing powers of h (see [Dr1] and references therein).

3.7.b. Yang–Baxter Equations with Parameters. The initial discovery of Yang–Baxter operators was connected with one-dimensional quantum mechanics and led to a slightly different algebraic structure. Imagine a system of vector spaces $V(t)$ depending on a parameter t, e.g., fibers of a vector bundle over a space T. Suppose that an operator $R(t_1, t_2) \in GL(V(t_1) \otimes V(t_2))$ is given for generic points t_1, t_2 in such a way that

$$\begin{aligned}(R(t_1,t_2) \otimes \mathrm{id})(\mathrm{id} \otimes R(t_1,t_3))(R(t_2,t_3) \otimes \mathrm{id})\\ = (\mathrm{id} \otimes R(t_2,t_3))(R(t_1,t_3) \otimes \mathrm{id})(\mathrm{id} \otimes R(t_1,t_2)).\end{aligned}$$

This means that the two ways to rearrange $V(t_1) \otimes V(t_2) \otimes V(t_3)$ in reverse order with the help of R coincide. Physically, R may correspond to the scattering operator of two particles moving with momenta $t_1, t_2; V(t_1)$ is an inner state space; momenta are conserved after the scattering.

One usually identifies all $V(t)$ (for example, by trivializing the vector bundle). A new feature of the situation is that one can now consider, for example, a meromorphic dependence of R on t_1, t_2; an important class of constructions leads to solutions with a pole at $t_1 = t_2$ so that one cannot rearrange fibers at the same point but only at different ones. Belavin and Drinfeld gave a classification of an important class of such solutions parametrized by an algebraic curve and having a pole of the first order on the diagonal.

They lead to bialgebras generated by families of multiplicative matrices $Z(t)$, with the commutation relations

$$R(t_1,t_2)Z(t_1) \odot Z(t_2) = Z(t_1) \odot Z(t_2)R(t_1,t_2),$$

which play an important role in two-dimensional physics.

3.7.c. Yang–Baxter Categories. A natural generalization of a Yang–Baxter operator and simultaneously a version of Yang–Baxter equations with parameters is given by the notion of a category with tensor product on tensor powers of objects on which a functorial action of braid groups is defined. We shall discuss this in some detail in the next section.

3.8. AN EXAMPLE: QUANTUM GL(2). As in the classical context, quantum GL(2) is a basic example and a germ of practically all aspects of the general theory.

3.8.a. Bialgebra $M_q(2)$. By definition, it is generated as an algebra by the coefficients of a matrix $Z = \begin{pmatrix} a & b \\ c & d \end{pmatrix}$ subject to the commutation relations

$$\begin{aligned} &ab = q^{-1}ba; \quad ac = q^{-1}ca; \quad bd = q^{-1}db; \quad cd = q^{-1}dc; \\ &bc = cb; \quad ad - da = (q^{-1} - q)bc. \end{aligned} \tag{3.5}$$

Here $q \in k^*$ is an arbitrary parameter.

This algebra has a bialgebra structure uniquely defined by the condition that Z is multiplicative. A direct proof is possible but cumbersome. For a conceptual proof of this and other properties, see [Ma2,4], and Section 4.2

Note that $(M_q(2), m^{\mathrm{op}}\Delta^{\mathrm{op}})$ is isomorphic to $M_{q^{-1}}(2)$.

3.8.b. Quantum Determinant. Put

$$D = \mathrm{DET}_q(Z) = ad - q^{-1}bc = da - qcb. \tag{3.6}$$

This is a multiplicative element: $\Delta(D) = D \otimes D, \varepsilon(D) = I$. It commutes with a, b, c, d. Moreover, if q is not a root of unity, D generates the center of $M_q(2)$.

3.8.c. Adjugate Matrix. We have

$$(3.7)\quad \begin{pmatrix} d & qb \\ -q^{-1}c & a \end{pmatrix}\begin{pmatrix} a & b \\ c & d \end{pmatrix} = \begin{pmatrix} a & b \\ c & d \end{pmatrix}\begin{pmatrix} d & qb \\ -q^{-1}c & a \end{pmatrix} = \begin{pmatrix} D & O \\ O & D \end{pmatrix}$$

3.8.d. Hopf Algebra $\mathrm{GL}_q(2)$. By definition, it is $M_q(2)[D^{-1}]$ endowed with the obvious diagonal map and the antipodal map derived from Eq. (3.7):

$$i(a) = D^{-1}d; \quad i(b) = -qD^{-1}b;$$
$$i(c) = -q^{-1}D^{-1}c; \quad i(d) = D^{-1}a.$$

Notice that the coefficients of $Z^{-1} = i(Z)$ satisfy the commutation relations of $M_q^{-1}(2)$, i.e., of the opposite bialgebra, as it should be, by general principles. A somewhat mysterious property of $\mathrm{GL}_q(2)$ is the following generalization.

3.8.e. "One-Parametric Subgroup" Passing via Z. The coefficients of Z^n satisfy Eq. (3.5) q^n for all integer n.[1]

In particular, if q is a root of unity, $M_q(2)$ and $\mathrm{GL}_q(2)$ contain a large commutative subring generated by the coefficients of Z^n. In fact, they are even finitely generated as modules over their centers.

3.8.f. Comodules. Put $\mathrm{SL}_q(2) = \mathrm{GL}_q(2)/(D-1)$. This is a Hopf algebra. Its category of, say, left comodules is semisimple precisely when q is not a root of unity. Simple comodules are classified by highest weights, as in the classical theory.

3.8.g. The Yang–Baxter Operator. Relations $(3.5)_q$ can be written in the form $RZ \odot Z = Z \odot ZR$, where R is the Yang–Baxter operator

$$R = \begin{pmatrix} q^{-1} & 0 & 0 & 0 \\ 0 & 1 & 0 & 0 \\ 0 & q^{-1}-q & 1 & 0 \\ 0 & 0 & 0 & q^{-1} \end{pmatrix}.$$

In fact, they were first discovered in this way. In Chapter 4, we reproduce another interpretation given in [Ma2].

We finish here our brief introduction to quantum groups based upon the ideas that originated in the work of the Leningrad school, Drinfeld, and Jimbo.

---

[1]This was remarked by Yu. Kobyzev in 1986 after my talk at a seminar (unpublished), and rediscovered recently by E. Corrigan, D. B. Fairlie, P. Fletcher, R. Sasaki (Preprint DTP-89-29, University of Durham, July 1989) and S. Vokos, B. Zumino, J. Wess (Preprint LAPP-TH-253/89, June 1989).

We shall reconsider this subject in Chapter 4 from a different angle, introducing quantum groups as symmetry objects in noncommutative geometry. Some basic examples of the Yang–Baxter quantum groups will also be given there, since they appear quite naturally also in the noncommutative geometry framework. A reader who is interested primarily in quantum groups may turn directly to the last chapter.

4. Monoidal and Tensor Categories as a Unifying Machine

4.1. PHILOSOPHY OF TENSOR CATEGORIES. Algebraic geometry over a field k is built from the spectra of commutative k-algebras so that the category of affine k-schemes is simply opposite to that of commutative k-algebras. The latter category in turn can be defined in terms of the category of linear k-spaces and the tensor product in it. Its object is just a diagram of linear spaces $k \xrightarrow{\eta} A \xleftarrow{m} A \otimes A$ with properties stated in Section 3.2(a), together with $m \circ S_{(12)} = m$, and morphisms are linear maps compatible with m, η. This suggests the possibility of exploring algebraic geometries based upon commutative rings in categories different from those of linear spaces but with similar tensor properties. Algebraic supergeometry is the simplest realization of this idea. Since $S_{(12)}$ in the category of $\mathbb{Z}_2$-graded spaces is twisted by a sign, a supercommutative ring is generally a noncommutative one, but the basic superalgebra can be modelled upon the standard theory. In a way, this is just one more example of a general principle of mathematical logic that an axiom system is never a categorical one. However, we usually are interested in the "external" properties of a new model, while the theory based upon initial axioms guarantees nontriviality and serves as a pattern.

There is a general heuristic principle: *A tensor category is a category of representations of something (Tanaka–Krein philosophy)*. (The words "tensor category" are used rather vaguely here; see also the more detailed discussion below.)

If this "something" is a commutative ring (or a scheme), the corresponding algebraic geometry is just the usual relative algebraic geometry over a base scheme.

If it is a group G, we get the algebraic geometry of G-schemes, G-sheaves, and so on, i.e., symmetry becomes a part of structure data. The relevance of this viewpoint is well known in topology, gauge field theory, and other areas.

Finally, "something" in the Tanaka–Krein principle may be a noncommutative Hopf algebra, e.g., a quantum group. This class of geometries is largely unexplored.

In order to obtain not only affine schemes, we probably have to ask for the existence of large commutative subalgebras in our relative commutative algebras. This would guarantee localization and all sorts of completion. Anyway, this seems to be one of the main reasons for the existence not only of algebraic supergeometry but also of analytic and differentiable geometries.

Of course, tensor categories were studied for a lot of reasons that had nothing to do with noncommutative geometry. In order to balance our biased presentation, the reader is advised to consult the following sources:

(a) MacLane's pioneering work [McL];

(b) Saavedra's thesis [S], where foundations were laid down for a "linearization" of the category of algebraic varieties via Grothendieck's vision of motives;

(c) Deligne and Milne [DM], and Deligne's papers [D2,3].

A reader with an analytic background should probably start much earlier, i.e., with Grothendieck's nuclear spaces, where the original motivation was basically the desire to concoct categories of topological vector spaces with tensor properties close to those of finite-dimensional spaces.

We shall now proceed with a more detailed discussion.

4.2. VECTOR SPACES AS A CATEGORY. As basic examples, we shall consider two categories: $\mathcal{V}$ (vector space over a field k) and $\mathcal{F}$ (finite-dimensional vector spaces over k). We shall list first some standard constructions in them and then their properties. (A part of) these constructions will become data in the axiomatic description of various classes of categories; (a part of) these properties become axioms. We shall try to stress what properties are needed for what purposes and also to discuss for which natural (and supernatural) reasons they can fail.

Data (for $\mathcal{F}$):

(a) The bifunctor $\otimes : \mathcal{F} \times \mathcal{F} \to \mathcal{F}$ with a commutativity constraint

$$S_{X,Y} : X \otimes Y \to Y \otimes X$$

and an associativity constraint

$$A_{X,Y,Z} : X \otimes (Y \otimes Z) \to (X \otimes Y) \otimes Z.$$

They are functorial isomorphisms defined for all objects X, Y, Z and given for $\mathcal{F}$ explicitly by

$$S_{X,Y}(x \otimes y) = y \otimes x; \qquad A_{X,Y,Z}(x \otimes (y \otimes z)) = (x \otimes y) \otimes z$$

for all $x \in X, y \in Y, z \in Z$.

(b) The identity object k together with identity constraints $k \otimes X \xrightarrow{L_X} X \xleftarrow{R_X} X \otimes k$ that are functorial isomorphisms.

(c) A dual object $X^\vee$ for each object X, together with the morphisms $\mathrm{ev} : X \otimes X^\vee \to k$ and $c : k \to X^\vee \otimes X$(concretely, $X^\vee$ = linear functions on X, ev is the canonical pairing, $c(1) = \sum x^i \otimes x_i$ in dual bases).

(d) $\mathcal{F}$ is an abelian k-linear category.

Properties (of $\mathcal{F}$):

(A) The commutativity constraint satisfies $S_{Y,X} \circ S_{X,Y} = \mathrm{id}_{X \otimes Y}$.

This strong requirement is not satisfied in important categories connected with Yang–Baxter equations. If it is, S is called a *symmetry*.

(B) The associativity constraint satisfies an axiom of the type "tensor product of four objects does not depend on the bracket positions" that can formally be stated as the commutativity of all diagrams ("pentagons")

$$\begin{array}{ccccc} X \otimes (Y \otimes (Z \otimes U)) & \xrightarrow{A} & (X \otimes Y) \otimes (Z \otimes U) & \xrightarrow{A} & ((X \otimes Y) \otimes Z) \otimes U \\ \downarrow \mathrm{id} \otimes A & & & & \uparrow A \otimes \mathrm{id} \\ X \otimes ((Y \otimes Z) \otimes U) & & \xrightarrow{A} & & (X \otimes (Y \otimes Z)) \otimes U \end{array}$$

(C) The associativity and commutativity constraints together satisfy a compatibility condition. If one allows oneself to write $X \otimes Y \otimes Z$ without brackets, it means that

$$S_{X \otimes Y, Z} = [S_{X,Z} \otimes \mathrm{id}_Y] \circ [\mathrm{id}_X \otimes S_{Y,Z}].$$

Otherwise, it becomes a commutative diagram with six objects ("hexagon axiom").

In fact, there is one more compatibility condition,

$$S_{X, Y \otimes Z} = [\mathrm{id}_Y \otimes S_{X,Z}] \circ [S_{X,Y} \otimes \mathrm{id}_Z].$$

It follows from the first one when S is a symmetry, but in general should be postulated independently.

(D) Duality axioms mean that the following two composed morphisms are identity morphisms,

$$(\mathrm{ev} \otimes \mathrm{id}_X) \circ (\mathrm{id}_X \otimes c) = \mathrm{id}_X;$$
$$(\mathrm{id}_{X^\vee} \otimes \mathrm{ev}) \circ (c \otimes \mathrm{id}_{X^\vee}) = \mathrm{id}_{X^\vee}.$$

(E) Identity axiom requires all diagrams

$$\begin{array}{ccc} (X \otimes k) \otimes Y & \xrightarrow{A} & X \otimes (k \otimes Y) \\ & {\searrow}^{R \otimes \mathrm{id}} \quad {}^{\mathrm{id} \otimes L}{\swarrow} & \\ & X \otimes Y & \end{array}$$

to be commutative.

TERMINOLOGY. An abstract category $\mathcal{T}$, endowed with a bifunctor $\otimes$ satisfying an associativity constraint and an identity object 1 (replacing k in $\mathcal{F}$) with identity constraints satisfying axiom E is called a *monoidal category*. Examples: $(\mathcal{F}, \oplus, 0)$; $(\mathcal{V}, \otimes, k)$ and $(\mathcal{V}, \oplus, 0)$; modules over a commutative ring, etc. If, in addition, it is endowed with a symmetry and a compatible commutativity constraint, it is called a *symmetric monoidal category*. Deligne and Milne call such categories simply *tensor categories*; a tensor category is called *rigid* in [DM] if it has internal $\underline{\mathrm{Hom}}$; $X^\vee = \underline{\mathrm{Hom}}(X, 1)$ is the functorial reflexive duality functor; and $\underline{\mathrm{Hom}}$ are multiplicative with respect to the tensor product. A lot more terminology is used elsewhere (see [S], and others). We shall sometimes indiscriminately use the general name "tensor category."

COMPARISON WITH $\mathcal{V}$. If we extend our category including vector spaces of infinite dimension, we shall have trouble with duality: c will not exist anymore, and infinite dimensional spaces are not reflexive. One remedy is to proceed to linear topology and continuous morphisms (cf. [A], where this machinery is worked out for the sake of Hopf algebras duality). Even then the new duality is not very compatible with the tensor product; usually there appears a new tensor product such that duality interchanges it with the initial one (see [Barr] and [Ma2,4]).

This shows that the whole package of properties of $(\mathcal{F}, \otimes)$ is too much to ask for.

If we replace k by a commutative ring, then again duality has good properties only for a restricted class of modules, namely, projective ones.

4.3. CATEGORIES OF REPRESENTATIONS. In Sections 3.4–3.6, we have discussed the category of finite-dimensional comodules over a bialgebra (or Hopf algebra) E and explained how one can introduce a tensor structure into this category.

For a change, we shall now do the same for a category of modules and explain also some variants of the notion of the Hopf structure, suggested by Drinfeld in [Dr1] and [Dr2].

Let E be a bialgebra over k, E-mod the category of left E-modules with the tensor product $X \otimes_k Y$, and the structure of the E-module on this product defined by inducing $E \longrightarrow E \otimes E$ via Δ. It is a monoidal category with identity object k (action via ε) and the standard associativity constraint induced by one in $\mathcal{V}$.

Drinfeld suggests the following method for introducing a commutativity constraint on E-mod. Suppose that one can find an invertible element $R \in E \otimes E$ such that for all $a \in E$, we have

(4.1) $\quad \Delta^{\mathrm{op}}(a) = R\Delta(a)R^{-1}.$

Denote by $R_{X,Y}$ the image of R in $\mathrm{End}_k(X \otimes Y)$ (where X, Y are two E-modules) and put

$$S_{X,Y} = S_{(12)} \circ R_{X,Y},$$

where $S_{(12)}$ is the usual transposition in $\mathcal{V}$. From Eq. (4.1), it follows that $S_{X,Y}$ forms a family of functorial isomorphisms $X \otimes Y \to Y \otimes X$.

Now the axioms take the form of the following restrictions on R.

(i) $S_{Y,X} \circ SX, Y = \mathrm{id}$ for all X, Y iff

(4.2) $\quad S_{(12)}(R) = R^{-1} \qquad$ *(unitarity condition)*

(ii) The two compatibility axioms in Section 4.2(C) are equivalent to

(4.3) $\quad (\Delta \otimes \mathrm{id})(R) = R_{13}R_{23}; \qquad (\mathrm{id} \otimes \Delta)(R) = R_{13}R_{12}$

(*quasitriangularity condition*). Here $R_{12} = R \otimes 1 \in E^{\otimes 3}$, and so on.

From Eq. (4.3), one can deduce the Yang–Baxter (or triangle) equation

(4.4) $\quad R_{12}R_{13}R_{23} = R_{23}R_{13}R_{12}.$

(iii) $S_{k,k} = \mathrm{id}$, iff

(4.5) $\quad (\varepsilon \otimes \varepsilon)(R) = 1.$

Drinfeld also uses the following terminology. A couple (E, R) (where R satisfies Eq. (4.1)) is called *quasitriangular* if Eq. (4.3) is true; *triangular* if Eqs. (4.2) and (4.3) are true; *coboundary* if Eqs. (4.2), (4.5), and

(4.6) $\quad R_{12} \cdot (\Delta \otimes \mathrm{id})(R) = R_{23} \cdot (\mathrm{id} \otimes \Delta)(R)$

are true. This condition means that the two ways to identify $X \otimes Y \otimes Z$ with $Z \otimes Y \otimes X$ coincide.

Finally, the duality in E-mod can be defined if E has a bijective antipode i: We put $X^\vee = \mathrm{Hom}_k(X, k)$ and define the action of $a \in E$ as the operator dual to $i^{-1}(a)$ acting on X.

A recent upsurge of interest in tensor categories for which $S_{X,Y}$ is *not* a symmetry has been connected with knot theory. As we have mentioned already, for an object X of such a category, the braid group Bd_n acts naturally on $X^{\otimes n}$, and one can build invariants of knots from characters of such representations. Quasitriangular and coboundary Hopf algebras (E, R) furnish a concrete approach to the construction of such categories.

The category E-mod can be considered as a category of vector spaces with an additional structure that allows us to define a twisting of the standard commutativity morphism $S_{(12)}$. It is natural to try to twist also the associativity morphism. In [Dr2], Drinfeld suggests to construct such categories as representations of *quasibialgebras* and *quasi-Hopf algebras* defined as follows.

4.4. DEFINITION. (a) A quasibialgebra is a quadruple $(A, \Delta, \varepsilon, \Phi)$, where A is an associative k-algebra with identity, $\Delta : A \to A \otimes A$ and $\varepsilon : A \to k$ are algebra homomorphisms, and Φ is an invertible element of $A^{\otimes 3}$ such that the following properties hold:

$$(4.7)\quad (\mathrm{id} \otimes \Delta)(\Delta(a)) = \Phi \cdot (\Delta \otimes \mathrm{id})(\Delta(a)) \cdot \Phi^{-1} \text{ for all } a \in A.$$

$$(4.8)\quad (\mathrm{id} \otimes \mathrm{id} \otimes \Delta)(\Phi) \cdot (\Delta \otimes \mathrm{id} \otimes \mathrm{id})(\Phi) = (1 \otimes \Phi) \cdot (\mathrm{id} \otimes \Delta \otimes \mathrm{id})(\Phi) \cdot (\Phi \otimes 1).$$

$$(4.9)\quad (\varepsilon \otimes \mathrm{id}) \circ \Delta = \mathrm{id} = (\mathrm{id} \otimes \varepsilon) \circ \Delta.$$

$$(4.10)\quad (\mathrm{id} \otimes \varepsilon \otimes \mathrm{id})(\Phi) = 1.$$

(b) A quasi-Hopf algebra is a quasibialgebra endowed with an antiautomorphism $i : A \to A$ and elements $\alpha, \beta \in A$ such that for any $a \in A$ with $\Delta(a) = \sum b_k \otimes c_k$, we have

$$(4.11)\quad \sum i(b_k) \alpha c_k = \varepsilon(a) \alpha; \qquad \sum b_k \beta i(c_k) = \varepsilon(a) \beta,$$

and, for $\Phi = \sum x_k \otimes y_k \otimes z_k$, $\Phi^{-1} = \sum p_k \otimes q_k \otimes r_k$,

$$(4.12)\quad \sum x_k \beta i(y_k) \alpha z_k = 1,$$

$$(4.13)\quad \sum i(p_k) \alpha q_k \beta i(r_k) = 1.$$

For a quasialgebra A, one introduces the tensor product on the category of A-modules in the same way as before. However, the associativity constraint $A_{X,Y,Z}$ is now defined as the image of Φ in $\mathrm{End}_k(X \otimes Y \otimes Z)$. From Eq. (4.7), it follows that this is an isomorphism of A-modules; Eq. (4.9) gives the identity constraints; Eq. (4.8) supplies the pentagon axiom, while Eq. (4.10) does the same for the identity axiom. Finally, the "quasiantipode" i can be used for the construction of the duality functor.

Of course, it is by no means clear a priori that nontrivial objects of this kind do exist. In fact they do, and Drinfeld gives a classification theorem for "formal" quasi-Hopf algebras. In the framework of perturbation theory, such algebras over $k[[h]]$ can be constructed by starting with a pair $(\mathfrak{g}, t)$, where $\mathfrak{g}$ is a Lie algebra and t is an invariant element of $S^2(\mathfrak{g})$. Then $A = U(\mathfrak{g})$, and $\Phi \in A^{\otimes 3}[[h]]$ is given by a universal formal series.

4.5. NONCOMMUTATIVE GEOMETRY BASED ON A TENSOR CATEGORY. For a rigid tensor category $\mathcal{T}$, one can relativize all notions of algebra defined usually with the help of $\mathcal{F}$ and $\mathcal{V}$ (instead of $\mathcal{V}$, one can take the category of inductive systems in $\mathcal{T}$). In order to introduce "external" notions that cannot be described in terms of the axioms common to all tensor categories, one should look first at $K_0(\mathcal{T})$ and $\mathrm{Pic}(\mathcal{T})$. Here K_0 is the Grothendieck ring of $\mathcal{T}$, where the values of "$\mathcal{T}$-dimension" lie. Similarly, $\mathrm{Pic}(\mathcal{T})$ is the group of classes of invertible elements L of $\mathcal{T}$, those for which $L \otimes L^\vee \to 1$ is an isomorphism. The central elements of this group can be used to define analogs of the parity change functor Π that plays such an important role in supergeometry.

The second natural task is an understanding of the Lie theory and algebraic affine group theory in such a category (cf. [G1,2]).

CHAPTER 2

Supersymmetric Algebraic Curves

1. A Superextension of the Riemann Sphere

1.1. RIEMANN SPHERE. A Riemann sphere is the space of $\mathbb{C}$-points of a projective line $\mathbb{P}^1$. Its automorphism group is PGL(2), and we can identify GL(2) with the group CSp(2) of conformal automorphisms of a symplectic form of rank 2.

In this section we shall define and study Riemann superspheres $\mathbb{P}^{1|1}$ and $\mathbb{P}^{1|2}$ as homogeneous spaces of conformal symplectic supergroups CSpO(2|1) and CSpO(2|2) respectively. Our main result is Theorem 1.12 that describes the geometric structures invariant with respect to these supergroups and in turn determining them. Riemann superspheres furnish the simplest examples of algebraic supercurves (those of genus zero) and simultaneously a starting point for uniformization theory.

We use elements of supergeometry as they are described in [Ma1]. This is the reason for the notation SpO which should have been replaced by Sp from the start (as well as OSp by O).

1.2. SUPERTRANSPOSITION. Let $B = \begin{pmatrix} B_1 & B_2 \\ B_3 & B_4 \end{pmatrix}$ be a matrix with coefficients in a supercommutative ring written in the standard formal [Ma1]. This means that B_1 consists of even-even places while B_4 of odd-odd ones. Recall that supertransposition is defined by

$$B^{\mathrm{st}} = \begin{pmatrix} B_1^{\mathrm{t}} & B_3^{\mathrm{t}} \\ -B_2^{\mathrm{t}} & B_4^{\mathrm{t}} \end{pmatrix} \quad (B \text{ even}); \quad B^{\mathrm{st}} = \begin{pmatrix} B_1^{\mathrm{t}} & -B_3^{\mathrm{t}} \\ B_2^{\mathrm{t}} & B_4^{\mathrm{t}} \end{pmatrix} \quad (B \text{ odd}),$$

where t is the usual transposition.

With a natural normalization, B^{st} describes a morphism of free modules conjugate to one defined by B. Here are the main properties of st:

(a) $(\mathrm{st})^4 = \mathrm{id}$, $(\mathrm{st})^2 \neq \mathrm{id}$. This reflects the fact that the general tensor category isomorphism $V \to V^{**}$ involves a sign change in the category of superspaces;

(b) $(BC)^{\mathrm{st}} = (-1)^{\hat{B}\hat{C}} C^{\mathrm{st}} B^{\mathrm{st}}$, where $(-1)^{\hat{B}}$ means parity of B;

(c) $\mathrm{Ber}(B^{st}) = \mathrm{Ber}(B)$ if B is a square even matrix (cf. [Ma1], Ch. III, sections 2,3).

1.3. THE SUPERGROUP $\mathrm{SpO}(2m|n)$. We shall denote by $I = I_{2m|n}$ the standard matrix of a split supersymplectic form

$$I_{2m|n} = \begin{pmatrix} I_0 & 0 \\ 0 & I_1 \end{pmatrix}, \quad I_0 = \begin{pmatrix} 0 & E_m \\ -E_m & 0 \end{pmatrix}, \quad I_1 = \begin{pmatrix} -1 & 0 & \dots & 0 \\ 0 & 0 & & -E_k \\ \vdots & & & \\ 0 & -E_k & & 0 \end{pmatrix}$$

for $n = 2k+1$, and similarly for $n = 2k$, with the first line and column of I_1 deleted. Here E means an identity matrix.

We define $\mathrm{CSpO}(2m|n)$ (or, for short, $\mathrm{C}(2m|n)$) as an affine algebraic supergroup given by equations

$$(1.1) \qquad \begin{cases} B^{st} I_{2m|n} B = z(B) I_{2m|n}, \\ z(B) \text{ is an invertible even scalar.} \end{cases}$$

This means that the entries of a generic even square matrix B of format $2m|n$ verifying Eq. (1.1), together with $z(B)$ and $z(B)^{-1}$, generate the coordinate ring (Hopf superalgebra) of $\mathrm{C}(2m|n)$. We write

$$B = \begin{pmatrix} B_0 & \Gamma_0 \\ \Gamma_1 & B_1 \end{pmatrix}.$$

Let now Z, Z' be columns of format $2m|n$. Define the symplectic form

$$(1.2) \qquad \langle Z, Z' \rangle = Z^{st} I Z'.$$

Clearly Eq. (1.1) means that

$$\langle BZ, BZ' \rangle = z(B) \langle Z, Z' \rangle,$$

i.e., $z(B)$ is a conformal multiplier. In short, CSpO consists of linear maps preserving the pairing Eq. (1.2) up to a scalar. From Eq. (1.1) one sees that

$$(1.3) \qquad z(B)^{2m-n} = \mathrm{Ber}(B)^2.$$

1.4. PROPOSITION. *$\mathrm{C}(2m|n)$ is an algebraic affine supermanifold of dimension $[(2m+1)m + n(n-1)/2 + 1]|2mn$. Its reduced space is*

$$\mathrm{C}(2m|n)_{\mathrm{red}} = \mathrm{C}(2m|0) \times_{\mathrm{GL}(1)} \mathrm{C}(0|n) = \mathrm{CSp}(2m) \times_{\mathrm{GL}(1)} \mathrm{CO}(n),$$

where the fibered product is taken with respect to the character $z(\cdot)_{\mathrm{red}}$.

Proof. The first equation (1.1) means that

$$B_0^t I_0 B_0 + \Gamma_1^t I_1 \Gamma_1 = z(B) I_0, \tag{1.4}$$

$$B_1^t I_1 B_1 - \Gamma_0^t I_0 \Gamma_0 = z(B) I_1, \tag{1.5}$$

$$B_0^t I_0 \Gamma_0 + \Gamma_1^t I_1 B_1 = 0 \tag{1.6}$$

By setting $\Gamma_0 = 0$, $\Gamma_1 = 0$ in Eqs. (1.4), and (1.5), we get equations for $C(2m|n)_{\mathrm{red}}$ that identify it with the corresponding fibered product. This also determines the even dimension of $\mathrm{C}(2m|n)$.

Since $z(B)$ is invertible, from Eqs. (1.4) and (1.5) one sees that $B_0^t I_0$ and $I_1 B_1$ are invertible. Therefore, $2mn$ odd elements of either Γ_0 or Γ_1 form independent global odd coordinates. Actually, Eqs. (1.4) and (1.5) do not constrain values of Γ_0 (or Γ_1) because, given a point of $C(2m|n)_{\mathrm{red}}$ and Γ_0, one can always adjust B_0, B_1 by even nilpotent corrections in order to satisfy Eqs. (1.4), and (1.5).

We shall later use explicit equations for $\mathrm{C}(2|1)$ and $\mathrm{C}(2|2)$.

1.5. THE GROUP $\mathrm{C}(2|1)$. Generic point:

$$B = \begin{pmatrix} a & b & \gamma \\ c & d & \delta \\ \alpha & \beta & e \end{pmatrix}$$

Equations:

$$ad - bc - \alpha\beta = z(B), \tag{1.4$'$}$$

$$e^2 + 2\gamma\delta = z(B), \tag{1.5$'$}$$

$$\begin{cases} \alpha e = a\delta - c\gamma, \\ \beta e = b\delta - d\gamma. \end{cases} \tag{1.6$'$}$$

By multiplying the two equalities (1.6′), we get $\alpha\beta e^2 = (ad - bc)\gamma\delta$. In particular, $\alpha\beta\gamma\delta = 0$. Therefore, $\alpha\beta(e^2 + 2\gamma\delta) = \gamma\delta(ad - bc - \alpha\beta)$ so that from Eqs. (1.4′) and (1.5′), we obtain in addition:

$$\alpha\beta = \gamma\delta, \quad e^2 = ad - bc - 3\alpha\beta.$$

From Eqs. (1.3) and (1.5′), it follows that

$$\mathrm{Ber}(B)^2 = ad - bc - \alpha\beta = (e + \alpha\beta e^{-1})^2.$$

Looking at $\mathrm{C}(2|1)_{\mathrm{red}}$, one sees that $\mathrm{Ber}(B) = e + \alpha\beta e^{-1} = z(B)$.

$C(2|1)$ is an irreducible supermanifold of dimension $4|2$. Its generic point B multiplies by $z(B)$ the symplectic form $z_1 z_2' - z_2 z_1' - \zeta_1\zeta_1'$.

1.6. THE GROUP $C(2|2)$. Generic point:

$$B = \begin{pmatrix} a & b & \gamma_1 & \gamma_2 \\ c & d & \delta_1 & \delta_2 \\ \alpha_1 & \beta_1 & e_1 & e_2 \\ \alpha_2 & \beta_2 & e_3 & e_4 \end{pmatrix}$$

Equations:

$$(1.4'') \quad ad - bc - \alpha_1\beta_2 - \alpha_2\beta_1 = z(B),$$

$$(1.5'') \quad \begin{cases} e_1e_3 = -\gamma_1\delta_1, \quad e_2e_4 = -\gamma_2\delta_2, \\ e_1e_4 + e_2e_3 + \gamma_1\delta_2 + \gamma_2\delta_1 = z(B); \end{cases}$$

$$(1.6'') \quad \begin{pmatrix} -c & a \\ -d & b \end{pmatrix}\begin{pmatrix} \gamma_1 & \gamma_2 \\ \delta_1 & \delta_2 \end{pmatrix} = \begin{pmatrix} \alpha_2 & \alpha_1 \\ \beta_2 & \beta_1 \end{pmatrix}\begin{pmatrix} e_1 & e_2 \\ e_3 & e_4 \end{pmatrix}$$

From Eq. (1.3), it follows that $\mathrm{Ber}(B) = \pm 1$. These two values correspond to two connected components of $C(2|2)$: On the identity component, e_1, e_4 are invertible, e_2, e_3 are nilpotent, and vice versa on the other one.

A point B of $C(2|2)$ multiplies by $z(B)$ the form

$$z_1z_2' - z_2z_1' - \zeta_1\zeta_2' - \zeta_2\zeta_1'.$$

The dimension of $C(2|2)$ is $5|4$.

There is an element of order two in $C(2|2)$, whose nonvanishing entries are $a = d = 1$, $e_2 = e_3 + 1$. Denote by $C^{\mathrm{sym}}(2|2)$ its centralizer. As we shall see later, it is a natural reduced structure subgroup for SUSY_2-curves. It consists of matrices in $C(2|2)$ with additional constraints:

$$\alpha_1 = \alpha_2 = \alpha; \quad \beta_1 = \beta_2 = \beta; \quad \gamma_1 = \gamma_2 = \gamma; \quad \delta_1 = \delta_2 = \delta;$$
$$e_1 = e_4 = e; \quad e_2 = e_3 = f.$$

The equations (1.4″) take the form

$$(1.4''') \quad ad - bc - 2\alpha\beta = z(B);$$

$$(1.5''') \quad \begin{cases} ef = -\gamma\delta; \\ e^2 + f^2 + 2\gamma\delta = z(B); \end{cases}$$

$$(1.6''') \quad \begin{cases} \alpha(e+f) = a\delta - c\gamma; \\ \beta(e+f) = b\delta - d\gamma. \end{cases}$$

We leave it to the reader to further investigate this group.

1.7. Projective Groups. Even invertible scalar matrices belong to $C(2m|n)$. We shall denote by $PC(2m|n)$ the corresponding quotient and explain how to define it in the category of algebraic supergroups for $2m|n = 2|1$ and $2|2$.

Since for a matrix of format $2|1$ we have $\mathrm{Ber}(\mathrm{diag}(a)) = a$, $PC(2|1)$ is canonically identified with the affine algebraic supergroup $SC(2|1) \subset C(2|1)$ consisting of matrices B with $\mathrm{Ber}(B) = 1$. On the other hand, since $z(B) = \mathrm{Ber}(B)^2$, the group $SpO(2|1)$ leaving $\langle Z, Z'\rangle$ invariant has two connected components $\pm SC(2|1)$. Therefore,

$$\begin{aligned} PC(2|1) &= SC(2|1) = SpO(2|1)_0, \\ SpO(2|1) &= (\pm 1) \times SpO(2|1)_0. \end{aligned}$$

All these groups have dimension $3|2$. Moreover,

$$PC(2|1)_{\mathrm{red}} = SL(2).$$

We turn now to format $2|2$. Here $\mathrm{Ber}(\mathrm{diag}(a)) = 1$ so that $SC(2|2)$ still contains scalar matrices. However, $z(\mathrm{diag}(a)) = a^2$. Therefore we can impose the equation $z(B) = 1$ defining the closed subgroup $SpO(2|2) \subset C(2|2)$ and an exact sequence

$$1 \to (\pm 1) \to SpO(2|2) \to PC(2|2) \to 1.$$

We have $\dim PC(2|2) = 4|4$. The kernel (± 1) is contained in the identity component. Hence $PC(2|2)$, as well as $SpO(2|2)_{\mathrm{red}} = Sp(2) \times_{GL(1)} O(2)$ consists of two connected components. Finally,

$$SpO(2|2)_{\mathrm{red},0} = SL(2) \times GL(1).$$

1.8. Riemann Superspheres and the Cross-Ratio. We now consider a projective superspace $\mathbb{P}^{1|N}$ with homogeneous coordinates $Z^{\mathrm{st}} = (z_1, z_2; \zeta_1, \ldots, \zeta_N)$, $N = 1$ or 2.

Clearly, $C(2|N)$ acts upon it via $Z' = BZ$, with even scalar matrices acting as identity. The projective line $\mathbb{P}^{1|N}$ considered as a $PC(2|N)$-space via this action will be called a Riemann supersphere.

Let now Z, Z', W, W' be four points of $\mathbb{P}^{1|N}$ given by their homogeneous coordinates. The function

$$(Z, W, Z', W') = \frac{\langle Z, W\rangle\ \langle Z', W'\rangle}{\langle Z, Z'\rangle\ \langle W, W'\rangle}, \tag{1.7}$$

where $\langle Z, W\rangle$ is the symplectic form (1.2), enjoys the following properties.

(a) It is an even rational function on $(\mathbb{P}^{1|N})^4$, and is $C(2|N)$-invariant with respect to the diagonal action.

(b) $(Z,W,Z',W') = (W,Z,W',Z')$
$= (Z',W',Z,W) = (Z,Z',W,W')^{-1}$.

(c) $(Z,W,Z',W')_{\text{red}}$ is the classical cross-ratio on $(\mathbb{P}^1)^4$.

The existence of this superversion of cross-ratio is a primary motivation for the reduction of the structure group $PGL(2|N)$ to $PC(2|N)$. For odd invariants, see Sections 2.12 and 2.18.

1.9. STRUCTURE DISTRIBUTIONS. The groups $PC(2|N)$ preserve certain distributions on $\mathbb{P}^{1|N}$ and actually coincide with their automorphism groups (for $N = 1,2$).

In order to explain this, consider the following coordinate-free construction. Let T be a linear superspace of dimension $2|N$ endowed with a split supersymplectic form, $\mathbb{P} = P(T)$ the associated projective superspace of lines in T, $\mathbf{T} = T \otimes \mathcal{O}_{\mathbb{P}}$, $\mathcal{T}_{\mathbb{P}}$ the tangent sheaf of $\mathbb{P}$. We can construct the usual (Euler) exact sequence

$$(1.8) \qquad O \to \mathcal{O}_{\mathbb{P}}(-1) \to \mathbf{T} \to \mathcal{T}_{\mathbb{P}}(-1) \to 0.$$

The sheaf $\mathbf{T}$ has an $\mathcal{O}_{\mathbb{P}}$-bilinear symplectic form induced by one on T. With respect to this form, $\mathcal{O}_{\mathbb{P}}(-1)$ is isotropic, and its orthogonal complement $\mathcal{O}_{\mathbb{P}}(-1)^{\perp}$ is of rank $1|N$. For $N = 2$, $\mathcal{O}_{\mathbb{P}}(-1)^{\perp}$ contains exactly two maximal isotropic subsheaves of rank $1|1$, say, $\mathbf{T}'$ and $\mathbf{T}''$. Put

$$\mathcal{T}^1(-1) = \text{ image of } \mathcal{O}_{\mathbb{P}}(-1)^{\perp} \text{ in } \mathcal{T}_{\mathbb{P}}(-1) \text{ for } N = 1;$$
$$\mathcal{T}'(-1),\ \mathcal{T}''(-1) = \text{ images of } \mathbf{T}', \mathbf{T}'' \text{ in } \mathcal{T}_{\mathbb{P}}(-1) \text{ for } N = 2.$$

We shall show below that $\mathcal{T}^1$, resp. $\mathcal{T}'$, $\mathcal{T}''$ satisfy the conditions of the following general definition.

1.10. DEFINITION. Let X be a complex supermanifold of dimension $1|N$, $N = 1$ or 2. The following data are called a SUSY_N-*structure* on X.

(a) For $N = 1$: a locally free locally direct subsheaf $\mathcal{T}^1 \subset \mathcal{T}_X$ of rank $0|1$ such that the Frobenius form

$$(\mathcal{T}^1)^{\otimes 2} \to \mathcal{T}^0 := \mathcal{T}_X/\mathcal{T}^1 : t_1 \otimes t_2 \to [t_1, t_2] \mod \mathcal{T}^1$$

is an isomorphism. In other words, $\mathcal{T}^1$ is a maximally nonintegrable distribution of corank 1, which is called a contact structure in the pure even geometry.

(b) For $N = 2$: two locally free locally direct subsheaves $\mathcal{T}'$, $\mathcal{T}''$ in $\mathcal{T}_X$ of rank $0|1$, whose sum in $\mathcal{T}_X$ is direct and which satisfy the integrability conditions $[\mathcal{T}', \mathcal{T}'] \subset \mathcal{T}'$, $[\mathcal{T}'', \mathcal{T}''] \subset \mathcal{T}''$, and the Frobenius form,

$$\mathcal{T}' \otimes \mathcal{T}'' \to \mathcal{T}^0 := \mathcal{T}_X/(\mathcal{T}' \oplus \mathcal{T}'') : t' \otimes t'' \to [t', t''] \mod (\mathcal{T}' \oplus \mathcal{T}'')$$

is an isomorphism.

Note the similarity of this structure and that of simple ($d = 4, N = 1$) supergravity as it was described in the last section of [Ma1]. Of course, this is not accidental, because a SUSY-structure is in fact a holomorphic version of two-dimensional supergravity that physicists often describe in terms of real geometry, metrics, and spinors.

In practical computations, it is sometimes convenient to introduce $\mathcal{T}^1$ by a family of odd vector fields D_i defined on charts U_i of an atlas, such that $D_i, D_i^2 = [D_i, D_i]/2$ form a basis for $\mathcal{T}_X$ on U_i and $D_i = F_i^j D_j$ on $U_i \cap U_j$, with F_i^j being a family of invertible even superfunctions. Clearly, putting $\mathcal{T}^1|U_i = \mathcal{O}_X D_i$ then defines a SUSY_1-structure.

Similarly, one can introduce $\mathcal{T}', \mathcal{T}''$ by a family of pairs of odd vector fields $\{D_i', D_i''\}$ on U_i such that $\{D_i', D_i'', [D_i', D_i'']\}$ form a local basis of $\mathcal{T}_{X'}$ and moreover,

$$\begin{aligned} (D_i')^2 &= f_i' D_i'; & (D_i'')^2 &= f_i'' D_i''; \\ D_i' &= F_i'^j D_j'; & D_i'' &= F_i''^j D_j'' \text{ on } U_i \cap U_j . \end{aligned}$$

1.11. EXAMPLES. (a) Let $N = 1, Z = (z, \zeta)$, a local coordinate system on X. Put $D_Z = \frac{\partial}{\partial \zeta} + \zeta \frac{\partial}{\partial z}$. Since $D_Z^2 = \frac{\partial}{\partial z}$, the subsheaf $\mathcal{T}^1 = \mathcal{O} D_Z$ defines a local SUSY_1-structure . We shall say that Z is a coordinate system *compatible* with this structure . If Z, Z' are compatible with the same SUSY_1-structure, we can define the multiplier $F_Z^{Z'}$ by $D_Z = F_Z^{Z'} D_{Z'}$. It can be called the *semijacobian*.

(b) Similarly, for $N = 2$, a local coordinate system $Z = (z, \zeta', \zeta'')$ defines the vector fields

$$D_Z' = \frac{\partial}{\partial \zeta'} + \zeta'' \frac{\partial}{\partial z}, \qquad D_Z'' = \frac{\partial}{\partial \zeta''} + \zeta' \frac{\partial}{\partial z}.$$

We have $(D_Z')^2 = (D_Z'')^2 = 0$, $[D_Z', d_z''] = 2\frac{\partial}{\partial z}$, so that we get a local SUSY_2-structure. Z is again called *compatible* with it.

We can now state and prove the main result of this section.

1.12. THEOREM. *(a) The structure distributions on* $\mathrm{P}^{1|N} = P(T)$, *corresponding to a supersymplectic form on* T (cf. Section 1.9) *define a* SUSY_N-*structure* ($N = 1, 2$) .

(b) The automorphism supergroup of this SUSY_N*-structure coincides with* $\mathrm{PC}(2|N)$*. For* $N = 2$*, an automorphism interchanges* $\mathbf{T}'$ *and* $\mathbf{T}''$ *iff it does not belong to the identity component.*

Proof. (a) Consider the covering of $\mathrm{P}^{(1|N)}$ by two coordinate charts:

$$\begin{aligned} U_1 : Z &= (z, \zeta) \text{ for } N = 1; & Z &= (z, \zeta', \zeta'') \text{ for } N = 2; \\ z &= z_1 z_2^{-1};\ \zeta = \zeta_1 z_2^{-1}; & \zeta' &= \zeta_1 z_2^{-1},\ \zeta'' = \zeta_2 z_2^{-1}. \\ U_2 : \overline{Z} &= (\overline{z}, \overline{\zeta}) \text{ for } N = 1; & \overline{Z} &= (\overline{z}, \overline{\zeta}', \overline{\zeta}'') \text{ for } N = 2; \\ \overline{z} &= -\overline{z}_2 \overline{z}_1^{-1};\ \overline{\zeta} = \overline{\zeta}_1 \overline{z}_1^{-1}; & \overline{\zeta}' &= \overline{\zeta}_1 \overline{z}_1^{-1},\ \overline{\zeta}'' = \overline{\zeta}_2 \overline{z}_1^{-1}. \end{aligned}$$

(Bar does not mean complex conjugation here!). First, we check that $D_{Z'}$, $D_{\overline{Z}}$ for $N = 1$ give one and the same SUSY-structure on $U_1 \cap U_2$ and similarly for $N = 2$. In fact, $D_Z = -\overline{z} D_{\overline{Z}}$ on $U_1 \cap U_2$ for $N = 1$, since both derivations transform z to ζ, ζ to 1. Similarly, $D'_Z = -\overline{z} D'_{\overline{Z}}, D''_Z = -\overline{z} D''_{\overline{Z}}$.

Let us now identify the corresponding distributions with those described in Section 1.9. We shall consider the chart U_1; U_2 is treated similarly.

Multiplying Eq. (1.8) by $\mathcal{O}_{\mathrm{P}}(1)$, we get

$$0 \to \mathcal{O}_{\mathrm{P}} \to \mathbf{T}(1) \xrightarrow{\varphi} \mathcal{T}_{\mathrm{P}} \to 0.$$

Here $\mathbf{T}(1)$ is generated by its global sections $\sum \ell_i \frac{\partial}{\partial z_i} + \sum \lambda_j \frac{\partial}{\partial \zeta_j}$, where ℓ_i, λ_j are linear forms in homogeneous coordinates. The image of $1 \in \Gamma(\mathcal{O}_{\mathrm{P}})$ in $\mathbf{T}(1)$ is the Euler vector field corresponding to $\ell_i = z_i, \lambda_j = \zeta_j$; φ transforms it into vanishing vector field on P.

It is easy to lift $D_{Z'}, D'_{Z'}, D''_Z$ to the sections of $\mathbf{T}(1)$ on U_1:

$$\begin{aligned} N = 1 : \varphi\left(\zeta_1 \frac{\partial}{\partial z_1} + z_2 \frac{\partial}{\partial \zeta_1}\right) &= \frac{\partial}{\partial \zeta} + \zeta \frac{\partial}{\partial z} = D_Z; \\ N = 2 : \varphi\left(\zeta_2 \frac{\partial}{\partial z_1} + z_2 \frac{\partial}{\partial \zeta_1}\right) &= \frac{\partial}{\partial \zeta'} + \zeta \frac{\partial}{\partial z} = D_{Z'}; \\ \varphi\left(\zeta_1 \frac{\partial}{\partial z_1} + z_2 \frac{\partial}{\partial \zeta_2}\right) &= \frac{\partial}{\partial \zeta''} + \zeta'' \frac{\partial}{\partial z} = D_{Z''}. \end{aligned}$$

We want to convince ourselves that $D_Z \in \mathcal{T}^1, D'_Z \in \mathcal{T}'$, $D''_Z \in \mathcal{T}''$, i.e., that these derivations lifted to $\mathbf{T}(1)$ are orthogonal to the Euler operator with respect to our symplectic form. Actually, for $N = 1$, we have

$$(z_1, z_2, \zeta_1) I_{2|1} \begin{pmatrix} \zeta_1 \\ 0 \\ -z_2 \end{pmatrix} = -z_2 \zeta_1 + \zeta_1 z_2 = 0,$$

and similarly for $N = 2$.

(b) Now we shall calculate the automorphism supergroups of the described SUSY-structures and prove that they coincide with $PC(2|1)$. To this end, we shall represent a point of $PGL(2|N)$ (with values in a supercommutative ring) by a linear fractional transformation, say, on U_1. Then we shall derive equations on the coefficients of this transformation expressing the preservation of structure distributions and see that they are equivalent to those of SpO.

This reasoning takes for granted that the automorphisms of the SUSY-structure belong to $PGL(2|N)$. Actually, for $N \neq 2$, one can prove that $\mathrm{Aut}(\mathbb{P}^{1|N}) = PGL(2|N)$ by calculating $\mathrm{Pic}(\mathbb{P}^{1|N})$ and by checking that there exists a unique invertible sheaf $\mathcal{O}(1)$ that reduces to the standard sheaf on $\mathbb{P}_{red}$ and has a maximal number of sections. For $N = 2$, this is wrong, and $\mathrm{Aut}(\mathbb{P}^{1|2})$ is strictly larger than $PGL(2|N)$. However, a given $SUSY_2$-structure allows one to define the relevant "$\mathcal{O}(1)$" as $(\Pi\mathcal{T}')^{-1}$.

We could also avoid some of the computations but preferred to be quite explicit for a later use.

(b_1) *Case* $N = 1$. Make a linear change of homogeneous coordinates:

$$\begin{pmatrix} w_1 \\ w_2 \\ \nu_1 \end{pmatrix} = \begin{pmatrix} a & b & \gamma \\ c & d & \delta \\ \alpha & \beta & e \end{pmatrix} \begin{pmatrix} z_1 \\ z_2 \\ \zeta_1 \end{pmatrix}.$$

On U_1 it takes the form

$$(1.9) \qquad \begin{aligned} w = w_1 w_2^{-1} &= \frac{az + b + e\zeta}{cz + d + \delta\zeta} = \frac{K}{L}; \\ \nu = \nu_1 w_2^{-1} &= \frac{\alpha z + \beta + e\zeta}{cz + d + \delta\zeta} = \frac{\Lambda}{L}. \end{aligned}$$

Expressing D_Z in terms of $\frac{\partial}{\partial w}$, D_W, where $W = (w, \nu)$, we get

$$(1.10) \qquad D_Z = D_Z^{\nu} \cdot D_W + (D_Z w - D_Z \nu \cdot \nu) \frac{\partial}{\partial w}.$$

We see that Eq. (1.9) preserves the SUSY-structure defined by $D_{Z'}$ iff

$$(1.11) \qquad D_Z w - D_Z \nu \cdot \nu = 0.$$

Putting Eq. (1.9) into Eq. (1.11) and clearing the denominators, we find

$$(1.12) \qquad L \cdot D_Z K - D_Z L \cdot K = D_Z \Lambda \cdot L,$$

or

$$(1.13) \qquad \begin{aligned} &(cz + d + \delta\zeta)(-\gamma + a\zeta) - (-\delta + c\zeta)(az + b + \gamma\zeta) \\ &\quad = (e - \alpha\zeta)(az + \beta + e\zeta). \end{aligned}$$

Comparing the coefficients at $1, z, \zeta, z\zeta$ in Eq. (1.13), we obtain a system of equations that is equivalent to Eqs. (1.4′)–(1.6′) defining $C(2|1)$. Since we consider fractional linear transformations, we have in fact $\mathrm{PC}(2|1)$.

(b_2) *Case* $N = 2$. The calculations are similar but somewhat longer. A linear transforation of homogeneous coordinates by a matrix B as in 1.6 leads to a substitution on U_1

$$\begin{aligned} w &= w_1 w_2^{-1} = \frac{az + b + \gamma_1\zeta' + \gamma_2\zeta''}{cz + d + \delta_1\zeta' + \delta_2\zeta''} = \frac{K}{L};\\ \nu' &= \nu_1 w_2^{-1} = \frac{\alpha_1 z + \beta_1 + e_1\zeta' + e_2\zeta''}{cz + d + \delta_1\zeta' + \delta_2\zeta''} = \frac{\Lambda'}{L};\\ \nu'' &= \nu_2 w_2^{-1} = \frac{\alpha_2 z + \beta_2 + e_3\zeta' + e_4\zeta''}{cz + d + \delta_1\zeta' + \delta_2\zeta''} = \frac{\Lambda''}{L}. \end{aligned} \tag{1.9'}$$

We have

$$D'_Z = D'_Z\nu' \cdot D'_W + D'_Z\nu'' \cdot D''_W + (D'_Z w - D'_Z\nu' \cdot \nu'' - D'_Z\nu'' \cdot \nu')\frac{\partial}{\partial w}, \tag{1.10'}$$

and similarly for D''_Z.

Therefore, Eq. (1.9′) preserves $\mathcal{T}' \oplus \mathcal{T}''$ iff

$$Dw = D\nu' \cdot \nu'' + D\nu'' \cdot \nu' \tag{1.11'}$$

for $D = D'_Z, D''_Z$, or, in view of (1.9′),

$$L \cdot DK - DL \cdot K = D\Lambda' \cdot \Lambda'' + D\Lambda'' \cdot \Lambda'. \tag{1.12'}$$

Comparing coefficients, as above, we get equations equivalent to Eqs. (1.4″)–(1.6″).

It remains to check that if Eqs. (1.10′) are valid, then $\mathcal{T}'$, $\mathcal{T}''$ either are preserved separately or interchanged, i.e., that either $D'_Z\nu'' = D''_Z\nu' = 0$ (at the identity component), or $D'_Z\nu' = D''_Z\nu'' = 0$ (outside of it). This can be verified by straightforward but rather lengthy computations.

One can also argue that the pair $(\mathcal{T}', \mathcal{T}'')$ is in fact uniquely determined by the conditions of Section 1.9 so that it is preserved by any map conserving the orthogonality with respect to the symplectic form.

1.13. Automorphy Factors. In the theory of automorphic functions on the Riemann sphere, an important role is played by the automorphy factor $j_g(z)$: If $z' = g(z) = (az + b)/(cz + d)$, $g \in \mathrm{GL}(2)$, we define it by the relation $\partial_z = j_g(z)\partial_{z'}$, where $\partial_z = \frac{\partial}{\partial z}$, and derive that

$$j_g(z) = \det(g)(cz + d)^{-2}.$$

On superspheres, we should consider D_Z (resp., D'_Z, D''_Z) instead of ∂_z.

(a) $N = 1$. For $B \in C(2|1)$, define $F_B(Z) = F_Z^{BZ}$ by $D_Z = F_Z^{BZ} D_{BZ}$, where Z is a coordinate system compatible with a given SUSY-structure. From Eq. (1.10), one sees that for $\binom{w}{\nu} = BZ$, one has

$$F_B(Z) = D_Z \nu,$$

which, after a calculation which takes into account Eqs. (1.4′)–(1.6′), gives

$$(1.14)\qquad \begin{aligned} &F(Z) = \mathrm{Ber}(B)/(cz + d + \delta\zeta) \\ &\quad (= 1/(cz + d + \delta\zeta) \text{ for } B \in SC(2|1) = PC(2|1)). \end{aligned}$$

(b) $N = 2$. For $B \in C(2|2)_0$, we similarly put

$$D'_Z = F'_B(Z) D'_{BZ}, \qquad D''_Z = F''_B(Z) D''_{BZ},$$

and for B in the other component,

$$D'_Z = F^{12}_B(Z) D''_{BZ}, \qquad D''_Z = F^{21}_B(Z) D''_{BZ}.$$

We leave to the reader the pleasant exercise of calculating these things explicitly. Writing

$$F_B(Z) = \begin{pmatrix} F' & 0 \\ 0 & F'' \end{pmatrix}, \quad \text{resp.} \quad \begin{pmatrix} 0 & F^{12} \\ F^{21} & 0 \end{pmatrix},$$

we can state the usual cocycle property for $N = 1$ and 2 simultaneously:

$$F_{BC}(Z) = F_C(Z) F_B(CZ).$$

1.14. LOBACHEVSKY'S SUPERPLANE ($N = 1$). As is well known, the upper half-plane $H = \{z | \mathrm{Im}(z) > 0\}$ together with the group of its holomorphic automorphisms $PSL(2, \mathbb{R})$ is a model of the Lobachevsky plane.

Turning to differential supergeometry, we can extend all the standard constructions to the superplanes. We restrict ourselves to the case $N = 1$ (without serious reasons).

Put again $Z = (z, \zeta), z = x + iy, \zeta = \xi + i\eta$. Consider $(z, \zeta; \bar{z}, \bar{\zeta})$ as complex coordinates on a real-analytic (or smooth) linear superspace (here the bar denotes complex conjugation). The Lobachevsky superplane is an open submanifold $\mathrm{Im}(z) > 0$ with boundary $\mathrm{Im}(z) = 0$. We shall consider it as a homogeneous space over the obvious real form of $PC(2|1) = SC(2|1)$.

1.15. Invariants and Covariants.

1.15.a. Imaginary Part. The correct analog of $y = \mathrm{Im}(z)$ on H is

$$Y = \text{“Im”}(Z) = \mathrm{Im}(z - \zeta\overline{\zeta}/2) = (z - \overline{z} - \zeta\overline{\zeta})/2.$$

In fact, if $Z' = BZ, Y' = \mathrm{Im}(z' - \zeta'\overline{\zeta}'/2)$ and B is a real point of $\mathrm{C}(2|1)$, we get

$$Y' = |F_B(Z)|^2 Y,$$

in complete analogy with the formula

$$\mathrm{Im}((az+b)(cz+d)^{-1}) = |cz+d|^{-2}(ad-bc)\,\mathrm{Im}(z).$$

Proof. Let $Z = (z_1, z_2, \zeta_1)^{st}$ in homogeneous coordinates, and similarly for Z'. Then

$$\begin{aligned} Y' &= \frac{1}{2}\frac{z_1'\overline{z}_2' - \overline{z}_1' z_2' - \zeta_1'\overline{\zeta}_1'}{z_2'\overline{z}_2'} = \frac{1}{2}\frac{\langle Z', \overline{Z'}\rangle}{|z_2'|^2} \\ &= \frac{1}{2}\frac{\langle BZ, \overline{BZ}\rangle}{|z_2'|^2} = \frac{z(B)}{|z_2'/z_2|^2}\frac{1}{2}\frac{\langle Z, \overline{Z}\rangle}{|z_2|^2} = |F_B(Z)|^2 Y. \end{aligned}$$

(We used the fact that B is real, the formula $z(B) = \mathrm{Ber}(B)^2$, and Eq. (1.14)).

1.15.b. Laplace Operator. Since for any holomorphic function Φ we have $\overline{D}_{Z'}\Phi = 0$, and $D_{Z'} = F_B(Z)^{-1} D_Z$, it follows that the real points of $\mathrm{C}(2|1)$ preserve the operator

$$\Box = 2iY\overline{Y}D_Z D_{\overline{Z}}.$$

This is the superanalog of the invariant Laplace operator $y^2(\partial_x^2 + \partial_y^2)$.

1.15.c. Invariant Distance. The cross-ratio (1.7) together with our real structure allows us to define an invariant of a pair of points $Z = (z, \zeta), W = (w, \nu)$ in $H^{1|1}$:

$$R(Z, W) = \frac{1}{4}\frac{\langle Z, W\rangle\langle \overline{Z}, \overline{W}\rangle}{\langle Z, \overline{Z}\rangle\langle W, \overline{W}\rangle} = \frac{|z - w - \zeta\nu|^2}{\mathrm{Im}(z - \zeta\overline{\zeta}/2)\,\mathrm{Im}(w - \nu\overline{\nu}/2)}.$$

Recall that for $\zeta = \nu = 0$ (i.e., on the reduced half-plane), we have

$$R(z,0;w,0) = 2\,\mathrm{ch}(d(z,w)) - 2,$$

where $d(z,w)$ is the Lobachevsky distance. See [BarMFS] for the utilization of this invariant in the theory of Selberg's superzetafunction.

1.15.d. Invariant Volume. The Berezin volume form v corresponding to $(z,\bar{z};\zeta,\bar{\zeta})$ under a coordinate change is multiplied by the squared modulus of the Jacobian matrix (cf. [Ma1]). Calculating it for real points of $C(2|1)$, one easily sees that

$$d\mu = -Y^{-1}v$$

is invariant and positive.

2. SUSY-Families and Schottky Groups

2.1. SUSY-FAMILIES. We shall now relativize Definition 1.10. Let $\pi : X \to S$ be a smooth morphism of complex superspaces. As is usual in algebraic geometry, we consider it as a family of fibers of π parametrized by a base S. Superfunctions on the base are parameters, or "relative constants." The relative dimension of π is the rank of the sheaf of vertical tangent fields, which we denote by $\mathcal{T}_{X/S}$. We shall call π a family of (super)curves if its relative dimension is $1|N$.

A SUSY_N-structure on such a family is a locally free locally direct subsheaf $\mathcal{T}^1 \subset \mathcal{T}_{X/S} = \mathcal{T}$ of rank $0|N$, for which the Frobenius form

$$\varphi : \wedge^2\mathcal{T}^1 \to \mathcal{T}^0 := \mathcal{T}/\mathcal{T}^1, \qquad \varphi(t_1 \wedge t_2) = [t_2, t_2] \mod \mathcal{T}^1$$

is nondegenerate and split, i.e., it locally has an isotropic direct subsheaf of maximal possible rank k for $N = 2k$ or $2k+1$ (cf. [Ma1], Chapter V, Section 6.5).

A SUSY_N-family is a family of curves with a SUSY_N-structure.

2.2. SUSY_1-FAMILIES AND THETA-CHARACTERISTICS. We first consider a family $\pi_0 : X_0 \to S$ of relative dimension $1|0$ over a pure even base S (i.e., $\mathcal{O}_S = \mathcal{O}_{S,0}$). Choose a relative theta-characteristic of this family, i.e., a pair (I, α), where I is an invertible sheaf of rank $1|0$ on X_0, and $\alpha : I^{\otimes 2} \to \Omega^1_{X/S}$ is an isomorphism.

Define a family of $1|1$-curves $\pi : X \to S$:

$$X_{\mathrm{red}} = X_{0,\mathrm{red}}, \;\; \mathcal{O}_{X,0} = \mathcal{O}_{X_0}, \;\; \mathcal{O}_{X,1} = \Pi I.$$

(Recall that Π is the parity change functor.) We shall introduce a SUSY_1-structure on X by the following construction. Let Z be a relative local coordinate on X_0, and ζ a local section of ΠI such that $\alpha(\zeta\otimes\zeta) = dz$, where $d = d_{X/S}$ is the vertical differential. Put $Z = (z,\zeta)$ and $\mathcal{T}^1 = \mathcal{O}_X D_Z$. The sheaf $\mathcal{T}^1$ does not depend on arbitrary choices: If $Z' = (z',\zeta'), z' = f(z)$, then $\zeta' = a(z)\zeta$, where $a(z)^2 = \frac{\partial f}{\partial z}$, and hence $D_{Z'} = a(z)^{-1} D_Z$.

In the next proposition, we shall show, in particular, that for any SUSY_1-family $\pi : X \to S$, its reduction with respects to odd constants $X_{\mathrm{rd}} \to S_{\mathrm{rd}}$ can be obtained by means of this construction. The corresponding theta-characteristic will be called the structural one (for π).

2.3. PROPOSITION. *For any family $\pi_0 : X_0 \to S$ of relative dimension $1|0$, the above construction defines a bijection between the following data.*

(a) Relative theta-characteristics of π_0 up to equivalence: $(I,\alpha) \simeq (I',\alpha')$ iff there is an isomorphism $I \to I'$ transforming α into α'.

(b) SUSY_1-families over $S, \pi : X \to S$, for which $X_{\mathrm{red}} = X_{0,\mathrm{red}}$ and $\mathcal{O}_{X,0} = \mathcal{O}_{X_0}$, up to isomorphism identical on X_0.

Proof. We shall construct an inverse map *(b)* $\Rightarrow$ *(a)*. Given $\pi : X \to S$, put $\Pi I = \mathcal{O}_{X,1}$. It is an invertible sheaf of rank $0|1$ on $X_0 = (X_{\mathrm{red}}, \mathcal{O}_{X,0})$, since π_0 is a smooth morphism of relative dimension $1|1$. The sheaf of odd vector fields $\mathcal{T}_1^1$ belonging to $\mathcal{T}^1$ after restriction to X_0 becomes an invertible $\mathcal{O}_{X,0}$-module J, and derivation defines a nondegenerate $\mathcal{O}_{X_0}$-pairing $J \otimes \Pi I \to \mathcal{O}_{X,0'}$, which is an isomorphism. Consider the Frobenius isomorphism $\varphi : (\mathcal{T}^1)^{\otimes 2} \to \mathcal{T}^0$, restrict it to X_0, and dualize. We obtain an isomorphism $\alpha : I^{\otimes 2} \to \Omega^1_{X_0/S}$. The pair (I,α) is the theta-characteristic we are looking for. We leave the remaining details to the reader.

2.4. EXAMPLE: CLASSIFICATION OF COMPACT SUSY_1-CURVES. A curve is a family over a point. If X_0 is $1|0$-dimensional, its SUSY_1-extensions form a tensor under $H^1(X_0,\mathbb{Z}_2)$. In particular, $X_0 = \mathbb{P}^1$ has a unique SUSY_1-extension that was described in Section 1.

If X_0 is compact of genus $g \geq 1$, card $H^1(X_0,\mathbb{Z}_2) = 2^{2g}$. This set consists of $2^g(2^g+1)$ odd and $2^g(2^g-1)$ even theta-characteristics (parity is $h^0(I) \bmod 2$). Since $H^0(\mathcal{O}_X) = \mathbb{C} \oplus H^0(\Pi I)$, in the case $h^0(I) > 0$, we get nonconstant odd global functions on X. Since $\mathcal{O}^2_{X,1} = 0$, for $h^0(I) > 1$, we have an ugly phenomenon: $\operatorname{Spec} \pi_* \mathcal{O}_X$ is not a supermanifold.

2.5. SUSY_2-FAMILIES AND THETA-PAIRS. Let again $\pi_0 : X_0 \to S$ be a $1|0$-family over a pure even base. Call a *relative theta-pair* a pair of invertible sheaves I', I'' on X_0 and an isomorphism $\beta : I' \otimes I'' \to \Omega^1_{X_0/S}$. Given such an object, we can construct a family of $1|2$-curves $\pi : X \to S$:

$$X_{\mathrm{red}} = X_{0,\mathrm{red}}; \qquad \mathcal{O}_{X,0} = \mathcal{O}_{X_0} \oplus I' \otimes I''; \qquad \mathcal{O}_{X,1} = \Pi(I' \oplus I'').$$

Here $I' \otimes I''$ is an ideal with zero multiplication. Introduce on X a SUSY$_2$-structure in the following manner. Let z be a relative local coordinate on X_0, ζ' (resp. ζ'') local sections of I' (resp. I'') such that $\beta(\zeta' \otimes \zeta'') = dz$, $d = d_{X/S}$. As in Section 1, put $Z = (z, \zeta', \zeta'')$ and introduce the vertical odd tangent fields D'_Z, D''_Z by formulas from Section 1.11 (b). Set $\mathcal{T}' = \mathcal{O}_X D_Z$, $\mathcal{T}'' = \mathcal{O}_X D''_Z$, $\mathcal{T}^1 = \mathcal{T}' \oplus \mathcal{T}'' \subset \mathcal{T}_{X/S}$. These distributions do not depend on Z. In fact, let $W = (f(z), a'(z)\zeta', a''(z)\zeta'') = (w, \nu', \nu'')$. Since $\beta(\nu' \otimes \nu'') = dw$, we have

$$D'_W = \left(a' + \zeta''\zeta'\frac{\partial a'}{\partial z}\right)^{-1} D'_Z; \qquad D''_W = \left(a'' + \zeta''\zeta'\frac{\partial a''}{\partial z}\right)^{-1} D''_Z.$$

(To see it, apply both sides to w, ν', ν''.)

However, before we state an analog of Proposition 2.3, we must take into account a small subtlety. It may happen that the distributions $\mathcal{T}'$, $\mathcal{T}''$ cannot be distinguished globally. More precisely, let $\pi : X \to S$ be a SUSY-family, $\mathcal{T}^1$ its structure distribution of rank $0|2$, and $\varphi : \Lambda^2\mathcal{T}^1 \to \mathcal{T}^0$ its Frobenius form. Denote by $\widehat{X}$ the relative grassmannian of φ-isotropic $0|1$-subsheaves in $\mathcal{T}^1$. The natural projection $\lambda : \widehat{X} \to X$ is unramified of degree 2 (cf. [Ma1], Ch.5, Section 6). Let X be connected.

2.6. Definition. A SUSY$_2$-structure is called *nonorientable* if $\widehat{X}$ is connected.

Otherwise it is called *orientable*, and a marking of one component is called its *orientation*.

Since on $\widehat{X}$ the sheaf $\lambda^*(\mathcal{T}_{X/S})$ coincides with $\mathcal{T}_{\widehat{X}/S}$, and the tautological sheaf of the isotropic grassmannian is a maximal isotropic subsheaf, a choice of orientation is equivalent to a numeration of two maximal odd integrable distributions $\mathcal{T}'$, $\mathcal{T}''$. Moreover, the lifted SUSY$_2$-structure on $\widehat{X}$ is oriented canonically, and $\widehat{X}$ is endowed with an involution reversing orientations.

Deligne, in a letter to the author, dated 2 December 1987, made the following remark: The category of *oriented* SUSY$_2$-families (over a given base) is equivalent to the category of plain $1|1$-families without an additional structure. A direct functor is $X \to X/\mathcal{T}' = Y$ (recall that $\mathcal{T}'$ is integrable and odd, therefore there are no global obstructions to form $X/\mathcal{T}'$). An inverse functor is: $Y \to \mathrm{Gr}_S(0|1, \mathcal{T}_{Y/S})$, where Gr_S is a relative grassmannian.

The involution $X/\mathcal{T}' \to X/\mathcal{T}''$ restricted to the category of families of $\mathrm{P}^{1|1}$'s plays an important role in the Borel–Weyl theory for supergroups developed by Penkov and Skornyakov. It is used to interpret "reflections with respect to odd roots."

2.7. PROPOSITION. *For any family $\pi_0 : X_0 \to S$ of relative dimension $1|0$ over a pure even base, the construction described in 2.5 defines a bijection between following data:*

(a) Relative theta-pairs for Π_0 up to the equivalence: $(I', I'', \beta) \simeq (I'_1, I''_1, \beta_1)$ iff there exist isomorphisms $I' \to I'_1, I'' \to I''_1$ transforming β into β_1.

(b) Oriented SUSY_2-families over $S, \pi : X \to S$, with $X_{\mathrm{red}} = X_{0,\mathrm{red}}$ and $\mathcal{O}_{X,0} = \mathcal{O}_{X_0}(1)$ up to isomorphism identical on $X_0^{(1)}$. Here $X_0^{(1)}$ is the first relative neighborhood of the diagonal in $X_0 \times_S X_0$.

Data (b) are called a SUSY_2-extension of X_0.

We omit the proof.

2.8. REMARKS. (a) As in Section 2.2, for any SUSY_2-family $\pi : X \to S$, we can construct a SUSY_2-family on S_{rd} by reducing odd constants. This defines a structure theta-pair of π_{rd}.

An orientable SUSY_2-family is called a symmetric one if it is isomorphic to the same family with reverse orientation. The sheaves I', I'' of the structure theta-pair of a symmetric family are isomorphic theta-characteristics. The SUSY_2-structure on $\mathbb{P}^{1|2}$ described in Section 1 is symmetric. There exist also nonsymmetric extensions of $\mathbb{P}^1$, but they are not isomorphic to $\mathbb{P}^{1|2}$.

Deligne's remarks about SUSY_2-families quoted earlier shows that the symmetric SUSY_2-families with an orientation-reversing involution should play a special role. This justifies the introduction of $\mathrm{C}^{\mathrm{sym}}(2|2)$ in Section 1.6. Note that this group is closer to $\mathrm{C}(2|1)$ than to $\mathrm{C}(2|2)$. However, we shall not pursue this line of thought further.

(b) Summarizing Proposition 2.3 and 2.7 as classification theorems, we can construct the following SUSY-families over a standard covering of the coarse moduli space M_g (or, for that matter, an appropriate moduli stack).

Let $M'_g \to M_g$ be the covering of degree 2^{2g} parametrizing pairs (a curve, a theta-characteristic), $X_0 \to M'_g$ the universal curve, and I the universal theta-characteristic (actually, it exists only over a moduli stack because I possesses the automorphism -1). Denote by $X \to M'_g$ the SUSY_1-family corresponding to I. Recall that M'_g consists of two components corresponding to even/odd I.

Furthermore, let $J_{g,d} \to M_g$ classify invertible sheaves of degree d on the universal curve. Let $Y_0 \to J_{g,d}$ be the lifting of the universal curve. Denote by I' the universal sheaf on Y_0, put $I'' = (I')^{-1} \otimes \Omega^1_{Y_0/J_{g,d}}$, and construct a SUSY_2 -family Y, extension of Y_0 by (I', I'').

For $d = g - 1$, there is an involution-changing orientation on this family, and $M'_g \subset J_{g,g-1}$ consists of its fixed points. The restriction of Y to M'_g gives a symmetric SUSY_2-family.

The constructed families are universal in even directions (at least, at smooth base points) but do not take into account the deformations in odd directions. Le Brun and Rothstein in [LBR] constructed moduli spaces for $SUSY_1$-families as infinitesimal extensions of M_g' by using an odd version of the Kodaira–Spencer technique. A very thorough investigation of local moduli problems in analytic supergeometry is due to Vaintrob [V]. Deligne, in a letter to the author, dated 25 September 1987, constructed a compactification of the moduli stack of $SUSY_1$-curves by using his new definition of the $SUSY_1$-structure on a stable singular curve. In another letter, he proved a theorem on the (quasi)fuchsian uniformization.

In the remaining part of this section, we shall construct some SUSY-families over supermanifolds by extending to supergeometry the classical Schottky uniformization. It would be interesting to investigate its (uni)versality properties.

2.9. A REVIEW OF CLASSICAL SCHOTTKY UNIFORMIZATION. For more detailed information on the following definitions and statements, see [He] and references therein.

2.9.a. Hyperbolic Elements. An element $t \in PSL(2, \mathbb{C})$ is called hyperbolic (or loxodromic) if there are two different points $z^{\pm}(t) \in \mathbb{P}^1(\mathbb{C})$ such that

$$\lim_{n \to \pm\infty} t^n(z_0) = z^{\pm}(t)$$

for any $z_0 \neq z^{\pm}(t)$. That means that a lifting of t in $GL(2, \mathbb{C})$ has proper values with different moduli. Points $z^{\pm}(t)$ are fixed by t. Denote by $q(t)$ the proper value of t on the cotangent space to $z^{+}(t)$; it is called the *multiplier* of t. Clearly, $0 < |q(t)| < 1$ and $z^{+}(t) = z^{-}(t^{-1})$, $q(t) = q(t^{-1})$. Transforming $(z^{+}(t), z^{-}(t))$ by a fractional linear map into $(0, \infty)$, we see that t is conjugate to $z \to qz, q = q(t)$.

Summarizing, denote by $Hyp(2) \subset PSL(2)$ the space of hyperbolic elements, put $D = \{q \in \mathbb{C} | 0 < |q| < 1\}$, and let $\Delta \subset \mathbb{P}^1 \times \mathbb{P}^1$ be the diagonal. Then the map

$$(2.1) \qquad D \times (\mathbb{P}^1 \times \mathbb{P}^1 \setminus \Delta) \to Hyp(2) : (q(t),\ z^{+}(t),\ z^{-}(t)) \to t$$

is an isomorphism of complex manifolds.

2.9.b. Schottky Groups. Here we shall call a Schottky group of genus g a discrete subgroup $\Gamma \subset PSL(2, \mathbb{C})$, which is free with g generators, consisting of hyperbolic elements (except of identity), and such that the set

$$\Omega_\Gamma = \mathbb{P}^1(\mathbb{C}) \setminus \Sigma_\Gamma,$$

where Σ_Γ is the closure of $\{z^{\pm}(t) | t \in \Gamma \setminus \mathrm{id}\}$, is nonempty, and connected.

2.9.c. Schottky Uniformization. Let Γ be a Schottky group of genus g. Then the quotient space $\Gamma \backslash \Omega_\Gamma$ is a compact Riemann surface of genus g. Each compact surface admits a Schottky uniformization. Conjugate subgroups uniformize isomorphic curves (but the converse statement is far from true, even for genus one).

2.9.d. The Schottky Domain. We can describe a Schottky group Γ by a family of its free generators. By conjugation, any such family can be transformed into $(t_1, \dots, t_g)$ with $z^+(t_1) = 0$, $z^-(t_1) = \infty$, and $z(t_2)^+ = 1$, if $g \geq 2$. We shall say that such a family is *normalized*. Denote by S_g the set

$$S_g = \{(q_1, z_1^+ = 0, z_1^- = \infty; q_2, z_2^+ = 1, z_2^-; \dots ; q_{g'}, z_g^+, z_g^-)$$
$$\text{such that } t_i \text{ with } q(t_i) = q_i, z^\pm(t_i) = z^\pm \text{ are free generators of a Schottky group}\}.$$

Thus S_g forms an open subset in the closed submanifold of $\mathrm{Hyp}(2)^g$ defined by the equations $Z_1^+ = 0, z_1^- = \infty, z_2^+ = 1$. We shall call it the Schottky domain. Its dimension equals 1 for $g = 1$, $3g - 3$ for $g \geq 2$. A complete system of inequalities defining S_g is seemingly unknown even for $g = 2$.

Let the point $s \in S_g$ correspond to the group $\Gamma(s)$ and put

$$\Sigma(s) = \Sigma_{\Gamma(s)}, \qquad \Omega(s) = \Omega_{\Gamma(s)}.$$

2.9.e. The Schottky Family. Put

$$\Omega = \bigcup_{s \in S_g} \Omega(s) \subset \mathrm{P}^1 \times S_g.$$

A free group Γ_g freely generated by $\{T_1, \dots, T_g\}$ acts fiberwise upon $\Omega : T_i$ acts by $t_i(s)$ over $s \in S_g$. The quotient space $X = \Gamma_g \backslash \Omega \in S_g$ is called the Schottky family. There is a natural surjective morphism $S_g \in M_g$ that is a nonramified covering of infinite degree.

Now we shall extend all this to supergeometry.

2.10. THE SUPERMANIFOLD $\mathrm{Hyp}(2|1)$. We shall show that one can define, by analogy with 2.9(a), a notion of a hyperbolic point of $\mathrm{SC}(2|1)$ with values in a supercommutative ring, and to introduce the supermanifold of hyperbolic elements of dimension $3|2$.

First of all, let $Z = (z, \zeta)$ be a SUSY-compatible coordinate system on $U_1 \subset \mathrm{P}^{1|1}$ as in Section 1. We define the point O by $z = 0, \zeta = 0$; point ∞ by $-z^{-1} = 0, \zeta z^{-1} = 0$. Consider a fractional linear map B from $\mathrm{SC}(2|1)$:

$$z' = \frac{az + b + \gamma\zeta}{cz + d + \delta\zeta}, \qquad \zeta' = \frac{dz + \beta + e\zeta}{cz + d + \delta\zeta}.$$

If $(0,\infty)$ are fixed points of B, we find immediately that $b=c=0, \alpha=\beta 0$. Then it follows from Eqs. (1.4′)–(1.6′) that $\gamma=\delta=0, e^2=\mathrm{ad}$. Finally, $\mathrm{Ber}(B)=1$ implies $e=1$. Therefore, by putting $a=d^{-1}=r$, we get that the subgroup of SC(2|1) fixing 0 and ∞ is GL(1):

$$z'=r^2 z, \qquad \zeta'=r\zeta. \tag{2.2}$$

Naturally, $r=r(B)$ should be called the *supermultiplier* of B. Clearly, $q(B_{\mathrm{red}})=r(B)^2_{\mathrm{red}}$. If r is a superfunction on a base superspace, the condition $0<|r_{\mathrm{red}}|<1$ identifies the subspace over which 0 is the attracting point for $B: 0=Z^+(B)$. An invariant description of $r(B)$ is that B multiplies the fiber of $\mathcal{T}^1$ at 0 by $r(B)^{-1}$.

Consider now the action of SC(2|1) upon $\mathbb{P}^{1|1}\times\mathbb{P}^{1|1}\setminus\Delta$. To simplify matters slightly, we shall restrict our attention to $U_1\times U_1$. Let $(Z^+,Z^-)\in U_1\times U_1, Z^+_{\mathrm{red}}\neq Z^-_{\mathrm{red}}$. We first write equations $BZ^+=0, BZ^-=\infty$:

$$az+b+\gamma\zeta^+ = 0; \qquad cz^-+d+\gamma\zeta^-=0; \tag{2.3}$$

$$\alpha z^+ +\beta+e\zeta^+=0; \qquad \alpha z^- +\beta+e\zeta^-=0. \tag{2.4}$$

We shall show now that Eqs. (2.3), (2.4), and (1.4′)–(1.6′) allow us to determine B uniquely up to one even parameter corresponding to supermultiplier.

First of all, we find from Eq. (2.3):

$$\alpha=\frac{e(\zeta^- -\zeta^+)}{z^+-z^-}, \qquad \beta=\frac{e(\zeta^+z^- -\zeta^-z^+)}{z^+-z^-}.$$

Now, from $\mathrm{Ber}(B)=e+\alpha\beta e^{-1}$, we see that

$$e=1-\frac{(\zeta^- -\zeta^+)(\zeta^+z^- -\zeta^-z^+)}{(z^+-z^-)^2}=\frac{z^+-z^- -\zeta^+\zeta^-}{z^+-z^-},$$

and this defines e, α, and β.

Using now Eqs. (1.6′) and then Eqs. (1.4′), and (1.5′), we can express γ through a,b;

$$\begin{aligned}\gamma&=(ad-bc)^{-1}(b\alpha-a\beta)e=(ad-bc-\alpha\beta)^{-1}(b\alpha-a\beta)e\\&=(e^2+2\alpha\beta)^{-1}(b\alpha-a\beta)e=e^{-1}(b\alpha-a\beta)=b\alpha-a\beta.\end{aligned}$$

Let us now rewrite Eqs. (2.3), (1.4′), and (1.5′):

$$\begin{aligned}a(z^+-\beta\zeta^+)+b(1+a\zeta^+)&=0;\\ cz^-+d-a\beta\zeta^-+b\alpha\zeta^-&=0;\\ ad-bc-3\alpha\beta&=e^2.\end{aligned}$$

Considering a as a free parameter, corresponding to the supermultiplier, we can find b from the first equation, and c and d from the second and third. Finally, δ is expressed via the rest of the entries. We can now state our result in geometric terms. We have constructed a map of supermanifolds

$$(2.5) \qquad D \times (\mathbb{P}^{1|1} \times \mathbb{P}^{1|1} \setminus \Delta) \to \mathrm{SC}(2|1) : (r, Z^+, Z^-) \to B.$$

2.11. PROPOSITION. *(a) The map* (2.5) *is an open embedding of supermanifolds. Denote its image by* Hyp(2|1) *(supermanifold of hyperbolic elements of* SC(2|1)*).*

(b) The restriction of Eq. (2.5) *onto reduced spaces can be included in the commutative diagram*

$$\begin{array}{ccc} (r, z^+, z^-) \in D \times (\mathbb{P}^{1|1} \times \mathbb{P}^{1|1} \setminus \Delta) & \xrightarrow{(2.5)_{\mathrm{red}}} & \mathrm{SC}(2|1)_{\mathrm{red}} = \mathrm{SL}(2) \\ \downarrow \qquad\qquad\qquad \downarrow & & \downarrow \\ (r^2, z^+, z^-) \in D \times (\mathbb{P}^{1|1} \times \mathbb{P}^{1|1} \setminus \Delta) & \xrightarrow{(2.1)} & \mathrm{PSL}(2) = \mathrm{SL}(2)/(\pm 1) \end{array}$$

In particular, $\mathrm{Hyp}(2|1)_{\mathrm{red}}$ *is the inverse image of* Hyp(2).

Before turning to the Schottky superdomain, consider the problem of "normalization" of Schottky generators, as in Section 2.9.d. The normalization prescription was based upon the following fact: Any ordered triple of different points of $\mathbb{P}^1$ can be transformed by a unique fractional linear map into $(0, \infty, 1)$. In the $N = 1$ supergeometry, there is a small new effect.

2.12. PSEUDOINVARIANT OF A TRIPLE ($N = 1$). Let (Z_1, Z_2, Z_3) be three points of $\mathbb{P}^{1|1}$ with different reductions. Suppose that B transforms (Z_1, Z_2) into $(0, \infty)$. By changing B by means of Eq. (2.2), one can achieve equality $BZ_3 = (1, \zeta)$, where ζ is *defined up to a sign*, if our points take values in a ring where square roots exist. In this way, we obtain a new invariant

$$(2.6) \qquad (Z_1; Z_2; Z_3) = (\pm\zeta) \in A_1/\mathbb{Z}_2$$

of A-valued point-triples of $\mathbb{P}^{1|1}$ with a SUSY-structure. We list some of its properties:

(a) $(0; \infty; (1, \zeta)) = \pm\zeta$.

(b) $(BZ_1; BZ_2; BZ_3) = (Z_1; Z_2; Z_3)$.

(c) $(Z_{\sigma(1)}; Z_{\sigma(2)}; Z_{\sigma(3)}) = \varepsilon(\sigma)(Z_1; Z_2; Z_3)$, where $\varepsilon(\sigma)(\pm\zeta) = (\pm\zeta)$ for even σ and $(\pm i\zeta)$ for odd σ.

If A is endowed with a real structure , we can construct from Eq. (2.6) a real even nilpotent invariant $|(Z_1; Z_2; Z_3)|^2$.

2.13. THE SCHOTTKY SUPERDOMAIN $S_g^{(1)}$. As in Section 2.9.d, consider first a closed subspace in Hyp(2|1) defined by the normalization conditions, and then its open submanifold, which is fiberwise determined by the Schottky conditions (essentially on the reduced space):

$$S_g^{(1)} = \{ (r_1, Z_1^+ = 0, Z_1^- = \infty; r_2, Z_2^+ = (1, \zeta), Z^- 2; \ \dots \ ; r_g, Z_g^+, Z_g^-)$$
such that $B_{i,\mathrm{red}}$ with these fixed points and supermultipliers freely generate a Schottky group at each point of the reduced space. }

Clearly,

$$\dim S_g^{(1)} = \begin{cases} 1|0 & \text{for } g = 1, \\ 3g-3|3g-2 & \text{for } g \geq 2. \end{cases}$$

2.14. SCHOTTKY'S SUSY$_1$-FAMILY. Proposition 2.11 shows that there exists an unramified covering of degree 2^g with the Galois group $\mathbb{Z}_2^g : S_g^{(1)} \to S_g$. Set

$\Omega^{(1)}$ = the open subsupermanifold in $\mathbb{P}^{1|1} \times S_g^{(1)}$ whose reduction is the inverse image of $\Omega \subset S_g$ under the natural map $(\mathbb{P}^{1|1} \times S_g^{(1)})_{\mathrm{red}} \to \mathbb{P}^1 \times S_g$.

Clearly, the free group Γ_g acts freely upon $\Omega^{(1)}$. The corresponding quotient superspace is our Schottky SUSY$_1$-family, whose SUSY-structure is induced by the standard one on $\mathbb{P}^{1|1}$, which is SC(2|1)- and therefore Γ_g-invariant.

This family contains all curves with *even* theta-characteristics. It is unclear (to me) whether one can modify this construction in order to also be able to Schottky uniformize odd characteristics. I hope this family is locally versal, although I did not check it.

2.15. SUPERMANIFOLD Hyp(2|2). Consider now a SUSY$_2$-compatible affine coordinate system (z, ζ', ζ'') on $U_1 \subset \mathbb{P}^{1|2}$, as in Section 1. Define the point O by $z = 0, \zeta' = \zeta'' = 0$; point ∞ by $-z^{-1} = 0, \zeta' z^{-1} = \zeta'' z^{-1} = 0$. Consider $B \in$ PC(2|2) defined by a matrix in SpO(2|2) denoted exactly as in Eq. (1.9′).

Assuming that $(0, \infty)$ are fixed points for B, we obtain $b = c = 0$; $\alpha_1 = \beta_1 = \alpha_2 = \beta_2 = 0$, whence $\gamma_1 = \gamma_2 = \delta_1 = \delta_2 = 0$, in view of Eq. (1.6″). Furthermore, on the connected component containing the identity, we have $e_2 = e_3 = 0$; on the other one, $e_1 = e_4 = 0$.

The remaining equations follow from Eqs. (1.4″) and (1.5″):

$$ad = e_1 e_4 = 1, \quad \text{or} \quad ad = e_2 e_3 = 1.$$

Putting $a = d^{-1} = r$, $e_1 d^{-1} = r'$, $e_4 d^{-1} = r''$ (resp. $e_2 d^{-1} = r^{12}$, $e_3 d^{-1} = r^{21}$), we reduce B to the form

$$(2.7) \qquad w = r^2 z, \quad \nu' = r'\zeta', \quad \nu'' = r''\zeta''; \quad r'r'' = r^2;$$

or

$$(2.7') \qquad w = r^2 z, \quad \nu' = r^{12}\zeta'', \quad \nu'' = r^{21}\zeta'; \quad r^{12}r^{21} = r^2.$$

We shall call (r', r'') (resp. (r^{12}, r^{21})) *supermultipliers*.

In the following, we shall consider only the identity component $\mathrm{PC}(2|2)_0$. If r', r'' are functions on a base space, O will be attracting over $0 < |r'r''|_{\mathrm{red}} < 1$. Multipliers r' (resp. r'') coincide with the action of B on $\mathcal{T}'$ (resp. $\mathcal{T}''$) at ∞.

Repeating the analysis of 2.10, consider now the action of $\mathrm{PC}(2|2)_0$ on $\mathbb{P}^{1|2} \times \mathbb{P}^{1|2} \setminus \Delta$. Let (Z^+, Z^-) lie in $U_1 \times U_1$ and let $BZ^+ = 0, BZ^- = \infty$. Then

$$(2.8) \qquad az^+ + b + \gamma_1\zeta'^+ + \gamma_2\zeta''^+ = 0; \; cz^- + d + \delta_1\zeta'^- + \delta_2\zeta''^- = 0;$$

$$(2.9) \qquad d_1 z^+ + \beta_1 + e_1\zeta'^+ + e_2\zeta''^+ = 0; \; d_1 z^- + \beta_1 + e_1\zeta'^- + e_2\zeta''^- = 0;$$

$$(2.10) \qquad \alpha_2 z^+ + \beta_2 + e_3\zeta'^+ + e_4\zeta''^+ = 0; \; \alpha_2 z^- + \beta_2 + e_3\zeta'^- + e_4\zeta''^- = 0$$

Let us show that this defines B up to two even parameters: a, e_1, e_4 on the identity component; a, e_2, e_3 outside.

From Eqs. (2.9) and (2.10), we can express α_1, β_1 linearly through e_j with known coefficients. From Eq. (1.6″) we similarly obtain expressions for γ_j, δ_j quadratic in e_j. By then using Eq. (1.5″), $e_1e_3 = -\gamma_1\delta_1$, $e_2e_4 = -\gamma_2\delta_2$, we can express two dependent e_i's through independent ones (take into account that dependent ones are nilpotents so that nonlinearity can be coped with).

We did not use a so far. Now express b, d through a, c and the already calculated entries, using Eq. (2.8). Putting it into Eq. (1.4″), we get $a(-cz^- - \delta_1\zeta'^- - \delta_2\zeta''^-) + c(az^+ + \gamma_1\zeta'^+ + \gamma_2\zeta''^+) - \alpha_1\beta_2 - \alpha_2\beta_1 = 1$, which allows us to find c.

Finally, from $e_1e_4 + e_2e_3 + \gamma_1\delta_2 + \gamma_2\delta_1 = 1$, we can find one of the parameters e_1, e_4 (resp. e_2, e_3).

In the geometric language, put $G = \{(r', r'') | 0 < |r'r''| < 1\} \subset \mathbb{C}^2$. We have constructed a map

$$(2.11) \qquad G \times (\mathbb{P}^{1|2} \times \mathbb{P}^{1|2} \setminus \Delta) \to \mathrm{SpO}(2|2)_0$$

and sketched the proof of the following result.

2.16. PROPOSITION. *The map (2.11) is an open embedding of supermanifolds. Its image* Hyp(2|2) *is the manifold of hyperbolic elements of* $\mathrm{SpO}(2|2)_0$.

2.17. PSEUDOINVARIANT OF A TRIPLE ($N = 2$). Let (Z_1, Z_2, Z_3) be a triple of points of $\mathrm{P}^{1|2}$ with pairwise disjoint supports. Assume that $BZ_1 = 0$, $BZ_2 = \infty$, $B \in \mathrm{PC}(2|2)_0$. Adjusting B by Eq. (2.7), we can achieve $BZ_3 = (1, \zeta', \zeta'')$. Only ζ, ζ'' is uniquely defined, and we can put

$$((Z_1; Z_2; Z_3)) = \zeta'\zeta''.$$

We leave the investigation of the properties of this invariant and the remaining steps of the construction of Schottky's SUSY_2-families to the reader.

3. Automorphic Jacobi-Schottky Superfunctions

3.1. GENERAL REMARKS. When a group acts discretely on a (super)-space, there are classical ways to construct automorphic functions using Eisenstein–Poincaré series or Weierstrass products over elements of the group. We shall show examples of such constructions with Schottky groups, stressing the formal algebra and working with sums and products as if they were absolutely covergent. Such a convergence is known on appropriate subdomains of the Schottky spaces in the classical case, and one can derive it over superextensions of these domains. We largely imitate in this exposition the constructions of [MaD] where they were applied in the theory of p-adic uniformization.

For a related example of superanalytic constructions, see [BarMFS], where a Selberg's superzeta function is discussed. Multipliers of hyperbolic elements, constructed in Section 2, are essentially used in the definition of the superzeta.

3.2. JACOBI–WEIERSTRASS PRODUCTS. Let $w(z)$ be an even meromorphic superfunction on $\mathrm{P}^{1|N} \times S_g^{(N)}$ rational in Z with nonvanishing reduction. Let $\Gamma = \Gamma_g$ be the universal Schottky group acting upon $\Omega^{(N)} = \mathrm{P}^{1|N} \times S_g^{(N)}$. Put formally

$$W_{Z_0}(Z) = \prod_{T'' \in \Gamma} \frac{w(T''Z)}{w(T''Z_0)}. \tag{3.1}$$

where Z_0 is a (local) section $S_g^{(N)} \to \Omega^{(N)}$. We shall assume that the reduced divisor of w lies in $\Omega_{\mathrm{red}}^{(N)}$ and that the support of its Γ-orbit does not intersect $Z_{0,\mathrm{red}}$. In the domain of absolute convergence, we have for $T \in \Gamma$,

$$W_{Z_0}(TZ) = \mu_w(T) W_{Z_0}(Z). \tag{3.2}$$

where the automorphy factor $\mu_w(T)$ is a superfunction, lifted from S and independent of Z_0. In fact,

$$W_{Z_0}(TZ) = \prod_{T' \in \Gamma} \frac{w(T'TZ)}{W(T'Z_0)} = \prod_{T'' \in \Gamma} \frac{w(T''Z)}{w(T''T^{-1}Z_0)},$$

whence, dividing by Eq. (3.1),

$$\mu_w(T) = \prod_{T'' \in \Gamma} \frac{w(T''Z_0)}{w(T''T^{-1}Z_0)}. \tag{3.3}$$

It follows that $\mu_w(T)$ is independent of Z. Since $W_{Z_0}(Z) = W_Z(Z_0)^{-1}$, independence of Z_0 also follows.

Consider now two elements $T, T' \in \Gamma$ and put

$$w'(Z) = \frac{\langle Z, T'Z_1 \rangle}{\langle Z, Z_1 \rangle}, \qquad w(Z) = \frac{\langle Z, T^{-1}Z_0 \rangle}{\langle Z, Z_0 \rangle}, \tag{3.4}$$

where $\langle \, , \rangle$ is defined in Section 1 by means of homogeneous coordinates (hopefully, the ambiguity in the notation Z will entail no confusion). Put $\Gamma^{ab} = \Gamma/[\Gamma, \Gamma]$, $\overline{T} = T \mod [\Gamma, \Gamma]$.

3.3. PROPOSITION. *Put*

$$\langle \overline{T} | \overline{T}' \rangle = \mu_{w'}(T). \tag{3.5}$$

This is a well-defined nondegenerate symmetric pairing on Γ^{ab} with values in $\mathcal{O}_S^*$ (restricted to the subdomain of the Schottky space where Eq. (3.3) converges absolutely).

Proof. From Eqs. (3.3) and (3.4) we find

$$\begin{aligned} \mu_{w'}(T) &= \prod_{T'' \in \Gamma} \frac{w'(T''Z_0)}{w'(T''T^{-1}Z_0)} \\ &= \prod_{T'' \in \Gamma} \frac{\langle T''Z_0, T'Z_1 \rangle}{\langle T''Z_0, Z_1 \rangle} \frac{\langle T''T^{-1}Z_0, Z_1 \rangle}{\langle T''T^{-1}Z_0, T'Z_1 \rangle} \\ &= \prod_{T'' \in \Gamma} (T''Z_0, T'Z_1, Z_1, T''T^{-1}Z_0), \end{aligned} \tag{3.6}$$

where the last cross-ratio is defined by Eq. (1.7).

Similarly,

$$
\begin{aligned}
\mu_w(T'^{-1}) &= \prod_{T''\in\Gamma} (T''Z_1, T^{-1}Z_0, Z_0, T''T'Z_1) \\
&= \prod_{T''\in\Gamma} (Z_1, T''^{-1}T^{-1}Z_0, T''^{-1}Z_0, T'Z_1) \qquad (3.7)\\
&= \prod_{T''\in\Gamma} (Z_1, T''T^{-1}Z_0, T''Z_0, T'Z_1).
\end{aligned}
$$

By comparing Eqs. (3.6) and (3.7), we obtain

$$\mu_{w'}(T) = \mu_w(T'^{-1}).$$

Since the l.h.s. does not depend on Z_0, while the r.h.s. does not depend on Z_1, and both are independent of Z, they can depend only on T and T'. Temporarily denoting this common value by $\langle T|T'\rangle$, we can rewrite the last identity as

$$\langle T|T'\rangle = \langle T^{-1}|T'^{-1}\rangle. \qquad (3.8)$$

Now notice that $\langle T|T'\rangle$ is bimultiplicative. In fact, if we denote $w(Z)$ more informatively as $w_{T,Z_0}(Z)$, we have

$$
\begin{aligned}
w_{T_1T_2,Z_0}(Z) &= \frac{\langle Z, T_2^{-1}T_1^{-1}Z_0\rangle}{\langle Z, Z_0\rangle} \\
&= \frac{\langle Z, T_2^{-1}T_1^{-1}Z_0\rangle\langle Z, T_1^{-1}Z_0\rangle}{\langle Z, T_1^{-1}Z_0\rangle\langle Z, Z_0\rangle} = w_{T_2, T_1^{-1}(Z_0)}(Z)\, w_{T_1,Z_0}(Z),
\end{aligned}
$$

and since $\mu_w(T'^{-1})$ does not depend on the choice of the normalization point Z_0, we get

$$\mu_{w_{T_1T_2}}(T'^{-1}) = \mu_{w_{T_2}}(T'^{-1})\mu_{w_{T_1}}(T'^{-1}),$$

that is

$$\langle T_1T_2|T'\rangle = \langle T_1|T'\rangle\langle T_2|T'\rangle. \qquad (3.9)$$

In the same manner, one can prove multiplicativity in T'. Symmetry then follows from Eq. (3.8). Finally, since $\langle T|T'\rangle$ takes values in an Abelian group, this symbol can depend only on $\overline{T}, \overline{T}'$.

The nondegeneracy of this pairing follows from its comparison with the classical period matrix of the reduced family, which we shall briefly recall now.

3.4. THE RIEMANN MATRIX. Let $(T_1, \dots, T_g)$ be the marked generator system of the universal Schottky group. Put

$$e^{2\pi i t_{ab}} = \langle \overline{T}_a, \overline{T}_b \rangle.$$

The matrix of even superfunctions (t_{ab}) on a subdomain of S is the super-analog of the Riemann matrix of an algebraic curve X. The latter is usually defined by a choice of a symplectic basis $(a_i, b_j) \in H_1(X, \mathbb{Z})$ and a Riemann basis of the first kind differentials:

$$\int_{a_i} \omega_j = \delta_{ij}; \qquad \int_{b_i} \omega_j = t_{ij}; \quad i, j = 1, \dots, g.$$

It is therefore desirable to have formulas for $\langle \overline{T} | \overline{T}' \rangle$ that would not contain explicitly the "mute" parameters Z_0, Z_1. In order to write them, we shall first recall a simple group-theoretical fact.

3.5. LEMMA. *(a) Let $T, T' \in \Gamma$ be such that the $\overline{T}, T'$ generate in Γ^{ab} a free direct Abelian subgroup of rank 2. Denote by $\mathrm{C}(T|T')$ a complete set of representatives of $(T) \setminus \Gamma / (T')$ in Γ. Then every element of Γ can be uniquely written in the form $T^m R T'^n$, where $m, n \in \mathbb{Z}$ and $R \in \mathrm{C}(T|T')$.*

(b) Let $\overline{T}$ generate a free direct Abelian subgroup in Γ^{ab}. Denote by $\mathrm{C}_0(T|T)$ a complete system of representatives of $(T) \setminus \Gamma / (T)$ without identity coset. Then every element of $\Gamma \setminus \{\mathrm{id}\}$ can be uniquely written in the form $T^m R T^n$, where $m, n \in \mathbb{Z}$, $R \in \mathrm{C}_0(T|T)$.

3.6. LEMMA. *(a) Assume that the conditions of Lemma 5(a) are fulfilled. Then*

$$\langle \overline{T} | \overline{T}' \rangle = \prod_{R \in \mathrm{C}(T'|T)} (Z^+_{T''}, RZ^+_{T'} RZ^-_{T'} Z^-_{T'}),$$

where $Z^{\pm}_T$ are the fixed sections of T.

(b) Assume that the conditions of Lemma 5b are fulfilled. Then

$$\langle \overline{T} | \overline{T} \rangle = r^2(T) \prod_{R \in \mathrm{C}_0(T|T)} (Z^+_{T'} RZ^+_{T'} RZ^-_{T'} Z^-_{T}),$$

where $r^2(T)$ is the supermultiplier squared for $N = 1$ (or $r'r''(T)$ for $N = 2$).

Proof. We start with Eq. (3.7):

$$\langle T|T'\rangle = \prod_{T''\in\Gamma} (Z_{1'}T''T^{-1}Z_{0'}T''Z_{0'}T'Z_1).$$

We collect together all elements of a double coset class $T'' = T'^{-m}RT^n$, $R \in C(T'|T)$, and multiply in the following order:

$$\langle T|T'\rangle = \prod_{R\in C(T'|T)} \prod_{m=-\infty}^{\infty} \prod_{n=-\infty}^{\infty} (Z_{1'}T'^{-m}RT^{n-1}Z_{0'}T'^{-m}RT^nZ_{0'}T'Z_1).$$

We shall first fix R, m and see how the product over n collapses. The nth term can be written as

$$\frac{\alpha_{n-1}}{\alpha_n}\frac{\beta_n}{\beta_{n-1}} = \frac{\langle Z_{1'}T'^{-m}RT^{n-1}Z_0\rangle}{\langle Z_{1'}T'^{-m}RT^nZ_0\rangle}\frac{\langle T'^{-m}RT^nZ_{0'}T'Z_1\rangle}{\langle T'^{-m}RT^{n-1}Z_{0'}T'Z_1\rangle},$$

whence

$$\prod_{n=-N}^{N}\frac{\alpha_{n-1}}{\alpha_n}\frac{\beta_n}{\beta_{n-1}} = \frac{\alpha_{-N-1}}{\alpha_n}\frac{\beta_N}{\beta_{-N-1}},$$

and, going to the limit $N \to \infty$.

$$\prod_{n=-\infty}^{\infty} = \frac{\alpha_{-\infty}\beta_\infty}{\alpha_\infty\beta_{-\infty}} = \frac{\langle Z_{1'}T'^{-m}RZ_T^-\rangle}{\langle Z_{1'}T'^{-m}RZ_T^+\rangle}\frac{\langle T'^{-m}RZ_{T'}^+T'Z_1\rangle}{\langle T'^{-m}RZ_{T'}^-T'Z_1\rangle}.$$

By multiplying all points by T'^m and by regrouping, we get

$$\prod_{n=-\infty}^{\infty} = \frac{\langle T'^mZ_{1'}RZ_T^-\rangle}{\langle T'^{m+1}Z_{1'}RZ_T^-\rangle}\frac{\langle T'^{m+1}Z,RZ_T^+\rangle}{\langle T'^mZ_{1'}RZ_T^+\rangle} = \frac{\gamma_m}{\gamma_{m+1}}\frac{\delta_{m+1}}{\delta_m}.$$

Now the same trick applies:

$$\prod_{m=-\infty}^{\infty}\prod_{n=-\infty}^{\infty} = \frac{\gamma_{-\infty}}{\gamma_\infty}\frac{\delta_\infty}{\delta_{-\infty}}$$

$$= \frac{\langle Z_{T'}^-RZ_T^-\rangle}{\langle Z_{T'}^+RZ_T^-\rangle}\frac{\langle Z_{T''}^+RZ_T^+\rangle}{\langle Z_{T''}^-RZ_T^+\rangle} = (Z_{T''}^+RZ_{T'}^+RZ_{T'}^-Z_{T'}^-).$$

(b) In the second statement of the theorem, all factors corresponding to $R \in C_0(T|T)$ are calculated in the same way as above and can be obtained by the formal substitution $T = T'$. In order to calculate the remaining terms, it suffices to modify slightly the very first computation by replacing $T'^{-m}R$ by id and T' by T. We get

$$\prod_{n=-\infty}^{\infty} (Z_1, T^{n-1}Z_0, T^n Z_0, T Z_1) = (Z_1, Z_T^-, Z_T^+, T Z_1).$$

It remains to calculate this cross-ratio in coordinates, where

$$Z_T^+ = 0,\ Z_T^- = \infty,\ Z_1 = (z, \zeta) \quad (\text{resp. } (z, \zeta', \zeta'')).$$

We get

$$T Z_1 = (r^2 z, r\zeta) \quad (\text{resp. } (r^2 z, r'\zeta', r''\zeta'')),$$

and furthermore,

$$\frac{\langle Z_1, 0\rangle}{\langle Z_1, \infty\rangle} = z, \qquad \frac{\langle T Z_1, 0\rangle}{\langle T Z_1, \infty\rangle} = r^2(T)z \qquad (\text{resp. } r'r''(T)z).$$

We finally note that in this coordinate system, we can similarly calculate $\langle \overline{T}|\overline{T}\rangle$:

$$(0, RZ_T^+, RZ_T^-, \infty) = \frac{\langle RZ_T^+, 0\rangle}{\langle RZ_T^+, \infty\rangle} : \frac{\langle RZ_T^-, 0\rangle}{\langle RZ_T^-, \infty\rangle},$$

$$\frac{\langle RZ_T^{\pm}, 0\rangle}{\langle RZ_T^{\pm}, \infty\rangle} = \text{the value of the even coordinate of } RZ \text{ at } Z = 0, \text{ resp. at } Z = \infty.$$

Therefore, for R as in 1.5 ($N = 1$) we have:

even coordinate of RZ at $0 = bd^{-1}$;
even coordinate of RZ at $\infty = ac^{-1}$;

and finally

$$\langle \overline{T}|\overline{T}\rangle = r^2(T) \prod_{R \in C_0(T|T)} bc/ad.$$

4. Superprojective Structures

4.1. DEFINITION. Let $\pi : X \to S$ be a SUSY_N-family. A *superprojective structure* on X is a (maximal) relative atlas, consisting of local coordinates Z on X/S, compatible with the SUSY-structure and such that its transition functions belong to $\mathrm{PC}(1|N)_0$.

Remark. An atlas compatible with a given SUSY-structure always exists. In order to prove that, it suffices to represent the reduced family by means of the structural theta-characteristics, to lift the resulting coordinate system locally and to correct it by nilpotent shifts (cf. [LBR], Lemma 1.2 for $N = 1$). Therefore the essential condition is imposed on transition functions.

Below we shall consider only the case $N = 1$.

4.2. DIFFERENTIAL OPERATORS. Denote by $\mathcal{D}$ the sheaf of the relative differential operators $\mathcal{O}_X \to \mathcal{O}_X$ over S. Since the Frobenius form is nondegenerate, $\mathcal{D}$ is generated (as an $\mathcal{O}_S$-algebra) by sections of $\mathcal{T}^1$ and multiplications by $\mathcal{O}_X$. In particular, if Z is a compatible local coordinate and $D = D_Z$, any section of $\mathcal{D}$ can locally be written as $K = \sum_{i=0}^{d} a_i D^i$. If $a_d \neq 0$, we define the superorder of K by $\mathrm{sord}_Z(K) = d/2$ (so that $\mathrm{sord}(\frac{\partial}{\partial z}) = 1$).

4.3. LEMMA. *Let Z, Z' be two local relative coordinate systems on X/S, $D = D_Z$, $D' = D_{Z'}$. Then the following statements are equivalent:*

(a) Z, Z' are compatible with the same (local) SUSY-*structure.*

(b) $D'z = \zeta D'\zeta$, where $(z, \zeta) = Z$.

(c) For some integer $i \geq 0$

$$\mathrm{sord}_Z(D'^{(2i+1)}) = (2i+1)/2 = \mathrm{sord}_Z(D^{2i+1}).$$

(d) Filtrations of $\mathcal{D}$ by sord_Z *and* $\mathrm{sord}_{Z'}$ *coincide.*

Proof. Clearly,

$$D' = D'\zeta \cdot D + (D'z - \zeta D'\zeta)D^2,$$
$$(D')^2 = D'^2\zeta \cdot D + (D'^2 z + \zeta D'^2\zeta)D^2.$$

It follows that (a)⇒(b)⇒(c) (for $i = 0$). One easily sees that (a)⇒(d) and (d)⇒(c). It remains to check that (c)⇒(b). We have

$$\begin{aligned}(D')^{2i+1} &= (D'^2)^i D' \\ &= [D'^2\zeta \cdot D + (D'^2 z + \zeta D'^2\zeta)D^2]^i [D'\zeta \cdot D + (D'z - \zeta D'\zeta)D^2].\end{aligned}$$

The highest power of D on the r.h.s. is D^{2i+2}. Its coefficient equals

$$A = (D'^2 z + \zeta D'^2\zeta)^i (D'z - \zeta D'\zeta).$$

But

$$(D'^2 z + \zeta D'^2 \zeta)_{\mathrm{red}} = \frac{\partial z_{\mathrm{red}}}{\partial z'_{\mathrm{red}}}$$

is invertible. Hence, (c)$\Rightarrow A = 0 \Rightarrow D'z - \zeta D'\zeta = 0$.

4.4. THE SCHWARZ SUPERDERIVATIVE. Recall that if Z, Z' are compatible with the same SUSY-structure, we can define the semijacobian $F = F_Z^{Z'}$ by $D = FD'$. Put

$$\sigma = \sigma_Z^{Z'} = \frac{D^3 F}{F} - \frac{2DF \cdot D^2 F}{F^2}. \tag{4.1}$$

This (odd) function on X is called *the Schwarz superderivative* (of Z' with respect to Z). In order to describe its properties, we put $\omega^i = (\mathcal{T}^1)^{\otimes(-i)}$ and denote by $L' = L_Z$ the differential operator

$$L' : \omega^{-1} \to \omega^2 : aD' \to D'^3 a \cdot (D')^{-2}.$$

(Here D'^3 is the result of the application of D'^3 to $a \in \mathcal{O}_X$, while $(D')^{-2}$ is a section of ω^2 and *not* an operator).

4.5. LEMMA. *We have*

$$L'(aD) = (D^3 a + \sigma_Z^{Z'} a) \cdot D^{-2} \tag{4.2}$$

Proof.

$$aD = (aF)(F^{-1}D) = aF \cdot D';$$
$$L'(aD) = D'^3(aF) \cdot (D')^{-2} = (F^{-1}D)^3(aF) \cdot F^2 \cdot D^{-2}.$$

4.6. COROLLARY. *We have*

$$\sigma_Z^{Z'} D^{-3} + \sigma_{Z'}^{Z''} (D')^{-3} = \sigma_Z^{Z''} (D)^{-3}.$$

Proof. According to the lemma,

$$L'(aD) = (D^3 a + \sigma_Z^{Z'} a) D^{-2}, \qquad L''(aD) = (D^3 a + \sigma_Z^{Z''} a) D^{-2}.$$

Put $aD = a'D'$, i.e., $a' = F_Z^{Z'} a$. On the one hand, we get

$$(L'' - L')(aD) = (\sigma_Z^{Z'} - \sigma_Z^{Z''}) a \cdot D^{-2}.$$

On the other hand, again by the lemma,

$$(L'' - L')(a'D') = (\sigma_{Z'}^{Z''} a')(D')^{-2} = (\sigma_{Z'}^{Z''} FaF^2) \cdot D^{-2}; \quad F = F_Z^{Z'}.$$

Hence

$$\sigma_Z^{Z'} - \sigma_Z^{Z''} = \sigma_{Z'}^{Z''} (F_Z^{Z'})^3.$$

Multiplying this by D^{-3}, we get (4.3).

4.7. PROPOSITION. *Let Z, Z' be relative local coordinates on X. The following statements are equivalent:*

(a) Z, Z' are compatible with a common SUSY*-structure, and $\sigma_Z^{Z'} = 0$.*

(b) There is a unique fractional linear transformation from $\mathrm{PC}(2|1) = \mathrm{SC}(2|1)$, *transforming Z into Z', with coefficients in $\mathcal{O}_S$. In other words, locally Z and Z' define a common superprojective structure.*

Proof. (a)⇒(b). Let $\sigma_Z^{Z'} = 0/$ We shall calculate the sheaf $\mathrm{Ker}(L')$ in coordinates Z and Z', taking into account that, in view of Eq. (4.2),

$$L'(aD') = (D'^3)a \cdot (D')^{-2}, \qquad L'(aD) = D^3 a \cdot D^{-2}.$$

Writing $\mathcal{O}_S$ instead of $\pi^{-1}(\mathcal{O}_S)$, we get

$$\mathrm{Ker}(L') = (\mathcal{O}_S \oplus \mathcal{O}_S z \oplus \mathcal{O}_S \zeta) \cdot D,$$
$$\mathrm{Ker}(L') = (\mathcal{O}_S \oplus \mathcal{O}_S z' \oplus \mathcal{O}_S \zeta') \cdot D', \qquad D' = F^{-1} D.$$

Therefore (perhaps, after a localization over S),

$$(4.4) \qquad \begin{pmatrix} z' \\ 1 \\ \zeta' \end{pmatrix} = \begin{pmatrix} a & b & \gamma \\ c & d & \delta \\ \alpha & \beta & e \end{pmatrix} \begin{pmatrix} F_z \\ F \\ F\zeta \end{pmatrix},$$

where the matrix B on the r.h.s. of Eq. (4.4) a priori belongs to $\mathrm{GL}(2|1, \mathcal{O}_S)$. It follows that

$$F = \frac{1}{cz + d + \delta\zeta}; \qquad z' = \frac{az + b + \gamma\zeta}{cz + d + \delta\zeta}; \qquad \zeta' = \frac{\alpha z + \beta + e\zeta}{cz + d + \delta\zeta}.$$

Since (z, ζ) and (z', ζ') are compatible with the same SUSY-structure, it follows from Theorem 1.12 that $B \in \mathrm{C}(2|1, \mathcal{O}_S)$. But, according to Eq. (1.14), we have

$$F = F_Z^{BZ} = \frac{\mathrm{Ber}(B)}{cz + d + \delta\zeta}.$$

Hence, $\mathrm{Ber}(B) = 1$, and $B \in \mathrm{SC}(2|1) = \mathrm{PC}(2|1)$.

(b)⇒(a). If Z, Z' are connected by a $B \in \mathrm{SC}(2|1)$, they define a common SUSY-structure, and Eq. (4.4) is fulfilled. Clearly, $\mathrm{Ker}(L')$ is generated by D', $z'D'$, $\zeta'D'$ over $\mathcal{O}_S$. From Eq. (4.4), it follows that it is generated also by D, zD, ζD. But these last sections generate also the kernel of $aD \to (D^3a) \cdot D^{-1}$. Therefore, it must be proportional to L'. Comparing symbols, one sees that it coincides with L', so that $\sigma_Z^{Z'} = 0$.

4.8. ASSOCIATED OPERATOR OF A SUPERPROJECTIVE STRUCTURE. Let X/S be a SUSY-family with a superprojective structure. In view of Proposition 4.7, we can globally define a differential operator

$$L : \omega^{-1} \to \omega^2, \qquad L(aD_Z) = D_Z^3 a \cdot D_Z^{-2},$$

where Z is an arbitrary coordinate system in the given atlas. We shall call this L the associated operator.

The previous discussion can be resumed in the following way.

4.9. THEOREM. *Let $\pi : X \to S$ be a* SUSY*-family admitting a superprojective structure. Then the set of all such structures can be embedded by the described construction into the set of operators $L : \omega^{-1} \to \omega^2$ which in any compatible coordinate system take form*

$$L(aD_Z) = (D_Z^3 + \sigma)D_Z, \quad \sigma \in (\omega^3)_1.$$

I thank a referee for the following remark. Locally on S superprojective structures do exist. To see it, one can apply the Cauchy theorem in supergeometry to $DF = D^2F = 0$ with the initial condition $F = 1$ at a chosen S-point, or, equivalently, to appeal to the classical result for the reduced family and then make nilpotent corrections.

It follows that the set of all local superprojective structures is a torsor over $\Gamma(S, \pi_*(\omega^3))_1$.

5. Sheaves of the Virasoro and Neveu–Schwarz Algebras

5.1. THE LIE ALGEBRA OF VIRASORO. This Lie algebra plays a fundamental role in some modern quantum-field-theoretical constructions since it is the common symmetry algebra of various two-dimensional conformal theories. It is a central extension of the Lie algebra of meromorphic vector fields on P^1 with singularities only at 0 and ∞:

$$V = \mathrm{C}[t, t^{-1}]\partial_t \oplus \mathrm{C}z; \qquad \partial_t = d/dt; \qquad z \text{ a central element};$$
$$[f\partial_t, g\partial_t] = (f\partial_t g - g\partial_t f)\partial_t + \frac{z}{12}\,\mathrm{res}(g\partial_t^3 f dt).$$

The element z or its eigenvalue at a representation of V is often called the *central charge*.

This construction can be generalized in various directions:

(a) $\mathbb{P}^1$ can be replaced by a Riemann surface or a SUSY-curve X of arbitrary genus, or else by a family of such curves.

(b) $\{0, \infty\}$ can be replaced by a finite family of points (or sections).

Since deleting extra points is consistent with the restriction of vector fields, it is appropriate to use the sheaf-theoretic language and construct a central extension of the sheaf of vector fields (vertical ones in the case of a family).

(c) The center of this extension is the de Rham cohomology sheaf $\Omega^1_X/d\,\mathcal{O}_X$ or its superversion.

Several constructions of this extension are now known. In this section, we shall explain such a construction based upon the consideration of the sheaf $\mathcal{D}$ and its noncommutative localization $\mathcal{E}$. In the subsequent sections, we shall exploit the Grothendieck–Sato realization of $\mathcal{D}$.

We shall consider in parallel the cases $N = 0$ (pure even families of curves) and $N = 1$ (SUSY_1-families).

5.2. Pseudodifferential Operators ($N = 0$). Let A be a commutative ring, $\partial : A \to A$ a derivation, and D the ring of differential operators generated by ∂ that consists of the formal polynomials $\sum_{i=0}^{n} a_i \partial^i$, $a_i \in A$ with the multiplication rule

$$(5.1)\qquad \left(\sum_i a_i \partial^i\right) \circ \left(\sum_j b_j \partial^j\right) = \sum_{i,j,k} \binom{i}{k} a_i \partial^k(b_j) \partial^{i+j-k}.$$

The ring of formal pseudodifferential operators E is obtained by formally inverting ∂ and completing. It consists of the formal series $\sum_{i=-\infty}^{n} a_i \partial^i$, $a_i \in A$, with the same multiplication rule (5.1), where the binomial coefficients are defined for all integers by $\binom{i}{k} = 0$ for $i < k$; 1 for $i = k$; $(k+1)\dots i/(i-k)!$ for $i > k$. For example,

$$\partial^{-1} \circ a = a\partial^{-1} - (\partial a)\partial^{-2} + (\partial^2 a)\partial^{-3} - \dots .$$

One can formally check that this defines a structure of an associative ring on E, which has the following properties.

(a) Put $E_n = \left\{ \sum_{i \le n} a_i \partial^i \right\}$. This is an ascending filtration. We have $E = D \oplus E_{-1}$; $E_i \circ E_j \subset E_{i+j}$; $[E_i, E_j] \subset E_{i+j-1}$.

(b) If $\partial' = a\partial$, a is invertible, and if D', E' are the similar rings generated by ∂' over A, then they can be canonically identified with D, resp. E, by

$\partial'^i \mapsto (a\partial)^{\circ i}$ for $i > 0$, or $(\partial^{-1} \circ z^{-1})^{\circ(-i)}$ for $i < 0$. This identification is also compatible with the filtration.

Using this remark, we can associate a sheaf of pseudodifferential operators with any one-dimensional distribution on a smooth or complex manifold X, using $\mathcal{O}_X$ in place of A and a local vector field generating the distribution in place of ∂.

(c) Put $\Omega = E_{-1}/E_{-2}$. This is a free A-module of rank 1. If $dz = 1$ for a certain $z \in A$, it is suggestive to write $\partial = \partial_z$ and $dz = \partial_z^{-1} \mod E_{-2}$, so that $\Omega = A dz$. The map $d : A \to \Omega$, $da = (\partial_z a) dz$ enjoys the usual standard properties of a differential. Sometimes we shall write $\nu = \nu_\partial$ instead of $\partial^{-1} \mod E_{-2}$.

(d) We have $E_i/E_{i-1} \simeq \Omega^{-i} = \Omega^{\otimes(-i)}$ for all i. The symbol map

$$\sigma_i : E_i \to \Omega^{-i}, \qquad \sigma_i\left(\sum_{j \leq i} a_j \partial^j\right) = a_i (\nu_\partial)^{-i}$$

does not depend on the choice of ∂ in the class $A^*\partial$.

(e) Let X be a compact Riemann surface, $\mathcal{E}$ the sheaf of formal pseudodifferential operators associated with $\mathcal{T}_{X'}$ and $\mathcal{E}_i$ the canonical filtration. Since $\mathcal{E}$ is a sheaf of Lie algebras, the associated graded sheaf $\oplus \mathcal{E}_i/\mathcal{E}_{i-1}$ is also a sheaf of Lie algebras. Therefore, we obtain a Lie algebra structure on the graded algebra of the pluricanonical embeddings $\bigoplus_{i \geq 0} \Gamma(\Omega_X^i)$.

For example, $[\omega_1, \omega_2] = \omega_2^2 \otimes d(\omega_1/\omega_2)$, if ω_1, ω_2 are two differentials of the first kind.

This Lie algebra deserves further investigation, as well as its nongraded version $\bigcup_i \Gamma(E_i)$.

We shall see below that the structure of the three consecutive layers $E_{i-3} \subset \ldots \subset E_i$ encodes the general Virasoro extension of the tangent sheaf.

5.3. Pseudodifferential Operators ($N = 1$). Now let A be a $\mathbb{Z}$-graded supercommutative ring and $D : A \to A$ an odd derivation:

$$(Da)\hat{} = \hat{a} + 1; \qquad D(ab) = Da \cdot b + (-1)^{\hat{a}} a \cdot Db.$$

We can repeat the constructions of the previous subsection with the following changes.

Instead of binomial coefficients, one should use the following superbinomial coefficients (cf. [MaR]):

$$\begin{bmatrix} i \\ k \end{bmatrix} = \begin{cases} 0 \text{ for } i < k \text{ and for } (i,k) \mod 2 = (0,1); \\ \begin{pmatrix} [i/2] \\ [k/2] \end{pmatrix} \text{ otherwise.} \end{cases}$$

Here $[i/2]$ means the integer part of $i/2$.

The ring of pseudodifferential operators E^s (s for super) consists of formal series in D with the multiplication rule

$$(5.2)\qquad (\Sigma a_i D^i) \circ (\Sigma b_j D^j) = \sum_{i,j,k} \begin{bmatrix} i \\ k \end{bmatrix} a_i D^k(b_j) D^{i+j-k}.$$

This rule is consistent with multiplication of differential operators. In order to compare E^s to E from Section 5.2, notice that $\partial = D^2$ is a derivation of the even subring A_0 of A. If we construct E by starting from (A_0, D^2), it embeds into E^s as $\{\sum a_{2i} D^{2i} | a_{2i} \in A_0\}$. Therefore, E^s is an extension of E by a square root of ∂ and odd superfunctions. This justifies the following definitions.

(a) Put $E^s_{n/2} = \left\{ \sum_{i \leq n} a_i D^i \right\}$. We have $E^s = D^s \oplus E^s_{-1/2}$, $E^s_i \circ E^s_j \subset E^s_{i+j}$.

(b) Let $D' = aD$, a be an invertible even element of A, and let E'^s be the corresponding ring. It can be canonically identified with E. Again, this will allow us to pass to sheaves (on supermanifolds with a given distribution of rank $0|1$, e.g., on SUSY_1-families).

(c) Put $\omega = E^s_{-1/2}/E^s_{-1}$. This is a free A-module of rank $0|1$ generated by $D^{-1} \mod E^s_{-1}$. If for certain $z \in A_0$, $\zeta \in A_1$, we have $Dz = \zeta$, $D\zeta = 1$, we denote $Z = (z, \zeta)$ and $D_Z = D$. We then write $dZ = D_Z^{-1} \mod e^s_{-1}$. The map $\delta : A \to \omega$ is well-defined, i.e., it remains the same when D is replaced by an element of $A_0^* D$. In general, we shall also write $v_D = D^{-1} \mod E^s_{-1}$, $\delta a = v_D \cdot Da \in \omega$.

(d) We have the symbol maps

$$\sigma_i : E^s_i \to E^s_i / E^s_{i-1/2} \simeq \omega^{2i}, \quad i \in \frac{1}{2}\mathbb{Z}.$$

(e) On a SUSY-curve (or a SUSY-family), one constructs sheaves $\mathcal{E}^s$ and $\mathcal{D}^s$ by localizing E^s and D^s.

5.4. Twisting. For $N = 0$, consider two A-modules L, M. Put $L^* = \underline{\mathrm{Hom}}(L, A)$. Set

$$E_{L \to M} = M \otimes_A E \otimes_A L^*$$

(note that E is an A-bimodule). Clearly, the subspace

$$D_{L \to M} = M \otimes_A D \otimes_A L^*$$

can be realized as a space of differential operators $L \to M$: for $m \in M$, $F \in D$, $\lambda \in L^*$, $l \in L$, we can put

$$(m \otimes F \otimes \lambda)l = m \cdot F\langle \lambda, l \rangle.$$

Of course, the whole of $E_{L\to M}$ does not act upon L in a natural way, but formally this space behaves as if it were a space of morphisms $L \to M$ in a category. In particular, there is a natural multiplication

$$E_{M\to N} \times E_{L\to M} \to E_{L\to N},$$

and $E_L := E_{L\to L}$ is a ring. The symbol map

$$\sigma_i : E_{i,L\to M} \to M \otimes \Omega^{-i} \otimes L^*$$

is defined in an evident way.

For $N = 1$, one can repeat these remarks with obvious changes. Modules are $\mathbb{Z}_2$-graded, Hom is the internal Hom.

All this is also compatible with sheafification.

FORMAL ADJUNCTION. There is a unique isomorphism of additive groups $E \to E_\Omega = \Omega \otimes E \otimes \Omega^{-1} : F \to F^t$, with the following properties:

(a) $f^t = v_\partial \otimes f \otimes v_d^{-1}$ for $f \in A$.

(b) $\partial^t = -v_\partial \otimes \partial \otimes v_\partial^{-1}$.

(c) $(F \circ G)^t = G^t \circ F^t$.

(d) This map is continuous in the ∂^{-1}-adic topology.

A key property is again the independence of t on the choice of ∂ in $A^*\partial$. An explicit formula allowing us to check this invariance is

$$\left(\sum a_i \partial^i\right)^t = v_\partial \otimes \left(\sum (-\partial)^i \circ a_i\right) \otimes v_\partial^{-1}.$$

For example, if $\partial' = a\partial$, we have

$$\begin{aligned} \text{in } E' : (\partial')^t &= -v_{\partial'} \otimes \partial' \otimes v_{\partial'}^{-1} = -v_\partial a^{-1} \otimes a\partial \otimes a v_\partial \\ &= -v_\partial \otimes (\partial \circ a) \otimes v_\partial^{-1}; \\ \text{in } E\ : (a\partial)^t &= \partial^t \circ a^t = -v_\partial \otimes (\partial \circ a) \otimes v_\partial^{-1}. \end{aligned}$$

Hence t extends to a morphism of sheaves.

Similarly, one can define a map

$$t : E_{L\to M} \to E_{M^t \to L^t}, \qquad L^t := L^* \otimes \Omega,$$

by the formula

$$(m \otimes F \otimes \lambda)^t = \lambda \otimes F^t \otimes m \in L^* \otimes \Omega \otimes E \otimes \Omega^{-1} \otimes M.$$

Here we must assume that L, M are free of finite rank.

If $L = M^t$ (and hence $M = L^t$), then t acts upon $E_{L \to M}$. In particular, t acts upon $\Omega^a \otimes E \otimes \Omega^{a-1}$. For each i, we can define a map

$$\sigma_i : \Omega^a \otimes E_i \otimes \Omega^{a-1} \to \Omega^a \otimes E_i/E_{i-1} \otimes \Omega^{a-1} = \Omega^{2a-1-i}.$$

Clearly, $\sigma_i(F^t) = (-1)^i \sigma_i(F)$. Hence, we can introduce i-selfconjugate operators by

$$E^+_{i,a} = \{F \in \Omega^a \otimes E_i \otimes \Omega^{a-1} | F^t = (-1)^i F\}.$$

A superversion of these constructions runs as follows. The conjugation map

$$\mathrm{t} : E^s \to E^s_\omega = \omega \otimes E^s \otimes \omega^{-1}, \quad F \to F^{\mathrm{t}}.$$

is defined by the properties

(a′) $f^{\mathrm{t}} = \nu_D \otimes f \otimes \nu_D^{-1}$ for $f \in A$.

(b′) $D^{\mathrm{t}} = -\nu_D \otimes D \otimes \nu_D^{-1}$.

(c′) $(F \circ G)^{\mathrm{t}} = (-1)^{\hat{F}\hat{G}} G^{\mathrm{t}} \circ F^{\mathrm{t}}$,

from which it follows that

$$\left(\sum a_i D^i\right)^{\mathrm{t}} = \nu_D \otimes \left(\sum (-1)^{i(i+1)/2 + i\hat{a}_i} D^i \circ a_i\right) \otimes \nu_D^{-1}.$$

Putting $L^{\mathrm{t}} = L^* \otimes \omega$, we can define

$$\mathrm{t} : E^s_{L \to M} \to E^s_{M^{\mathrm{t}} \to L^{\mathrm{t}}}$$

by

$$(m \otimes F \otimes \lambda)^{\mathrm{t}} = (-1)^{\hat{\lambda}\hat{F} + \hat{\lambda}\hat{m} + \hat{m}\hat{F}} \lambda \otimes F^{\mathrm{t}} \otimes m.$$

As in the even case, t acts upon $\omega^a \otimes E^s \otimes \omega^{a-1}$, and the symbol map

$$\sigma_i : \omega^a \otimes E^s_i \otimes \omega^{a-1} \to \omega^a \otimes E^s_i/E^s_{i-1/2} \otimes \omega^{a-1} = \omega^{2a-1-2i}$$

satisfies the sign rule

$$\sigma_i(F^{\mathrm{t}}) = \varepsilon_F \sigma_i(F); \qquad \varepsilon_F + (-1)^{(2i+1)\hat{F} + 2i(2i+1)}.$$

Using this, we define i-self-conjugate operators by

$$E^{s+}_{\iota,a} = \{F \in \omega^a \otimes E^s_\iota \otimes \omega^{a-1} \mid F^\iota = \varepsilon_F F\}.$$

5.5. THREE LAYERS OF THE FILTRATION. We can calculate explicitly the form of self-conjugate operators; by looking only at the first three coefficients, we get the following results. Put

$$\text{for } N=0: \quad \overline{V}_n = [\Omega^{n/2} \otimes E_n/E_{n-3} \otimes \Omega^{n/2-1}]^+;$$
$$\text{for } N=1: \quad \overline{K}_n = [\omega^{(n-1)/2} \otimes E^s_{n/2}/E^s_{(n-3)/2} \otimes \omega^{(n-3)/2}]_0 \quad \text{(even part)}.$$

Then, writing for clarity $\partial = \partial_z$, $D = D_Z$, we have:

5.6. LEMMA. *(a) $\overline{V}_n$ consists of the expressions*

$$F_{a,c} = (dz)^{n/2} \otimes \{a\partial^n + n\partial a \cdot \partial^{n-1}/2 + c\partial^{n-2} \mod (\partial^{n-3})\} \otimes (dz)^{(n-2)/2},$$

with arbitrary a, $c \in A$.

(b) $\overline{K}_n$ consists of the expressions

$$G_{a,c} = (dZ)^{(n-1)/2} \otimes \{aD^n + Da \cdot D^{n-1}/2 + (n-1)D^2a \cdot D^{n-2}/4 + cD^{n-3} \mod (D^{n-4})\} \otimes (dZ)^{(n-3)/2},$$

where $a \in A_0$, $c \in A_1$.

5.7. THE VIRASORO EXTENSION. By using Lemma 5.6(a), we can construct the following exact sequence, which does not depend on the choice of ∂ in $A^*\partial$:

$$(5.3) \qquad 0 \to \Omega \to V \xrightarrow{\sigma_n} \Omega^{-1} \to 0.$$

In fact, $\sigma_n(F_{a,c}) = a\partial$, so that $\mathrm{Ker}(\sigma_n) = \{F_{0,c}\}$ can be identified with Ω via $\sigma_{n-2} : F_{0,c} \to c\,dz$.

In order to make sense of $(dz)^{n/2}$, we have to assume n even.

We now set

$$V_n = \overline{V}_n/dA; \qquad dA \subset \Omega \subset \overline{V}_n; \qquad H = \Omega/dA.$$

From Eq. (5.3) we obtain an exact sequence

$$(5.4) \qquad 0 \to H \to V_n \to \Omega^{-1} = T \to 0.$$

5.8. THEOREM. *V_n is endowed with the natural structure of a Lie algebra, H belongs to its center, and σ_n is a homomorphism of Lie algebras. This extension corresponds to the cocycle*

$$c(b\partial_z, a\partial_z) = \frac{1}{12}n(n-1)(n-2)a\partial_z^3(b)dz \mod dA,$$

which is a direct generalization of the Virasoro cocycle.

Proof. The Lie algebra $T = \Omega^{-1} = A\partial \subset E$ acts on all natural A-modules (Lie derivative). Namely, T acts on E via adjoint representation: $[X, F] - X \circ F - F \circ X$ for $X \in T$, $F \in E$. Since $[T, E_i] \subseteq E_i$, this induces an action on $\Omega^i = E_{-i}/E_{-i-1}$. Finally, by using the Leibniz rule, we extend this action to the tensor product of these modules. In particular, from Eq. (5.1), we get

$$\mathrm{Lie}_{\partial_z}(dz^m) = [b\partial_z, dz^m] = (b\partial_z a + ma\partial_z b)(dz)^m.$$

Therefore,

$$[b\partial_z, F_{a,c}] = F_{a',c'},$$

where

$$a' = b\partial_z a - a\partial_z b; \qquad c' = n(n-1)(n-2)a\partial_z^3 b/12 + \partial_z(bc).$$

This means that $\overline{V_n}$ is a T-module, σ_n is a morphism of T-modules, and the action of T maps $\Omega \subset \omega V_n$ into dA (in our notation, Ω corresponds to $a = 0$). Hence, the induced action on $V_n = \overline{V_n}/dA$ is trivial on H, and Eq. (5.4) becomes a central extension of T.

5.9. THE NEVEU-SCHWARZ SUPERALGEBRA AND THE CONTACT LIE ALGEBRA. We now turn to the odd derivation case. First, recall that the Neveu-Schwarz Lie superalgebra NS is traditionally defined by even generators e_i, $i \in \mathbb{Z}$; odd generators f_α, $\alpha \in \mathbb{Z} + 1/2$; and the central even generator z, with the commutator rules

$$[e_i, e_j] = (j-i)e_{i+j} + \delta_{i,-j}(j^3 - j)z/12;$$
$$[e_i, f_\alpha] = (\alpha - i/2)f_{\alpha+i};$$
$$[f_\alpha, f_\beta] = 2e_{\alpha+\beta} + \delta_{\alpha,-\beta}(\alpha^2/3 - 1/12)z.$$

We put $\mathrm{NS}_\Lambda = \Lambda \otimes_{\mathbb{C}} \mathrm{NS}$ for an arbitrary supercommutative $\mathbb{C}$-algebra Λ. We shall explain how NS_L is related to the contact algebra.

The (central extension of the) contact Lie algebra K_L can be defined as $\Lambda_0 z \oplus A_0$, where Λ_0, A_0 are the even parts of Λ and $A = \Lambda[t, t^{-1}, \tau]$ respectively, $\hat{t} = 0$, $\hat{\tau} = 1$, with the commutation rule

$$\{a, b\} = aD^2 b - bD^2 a + DaDb/2 - \rho(aD^5 b)z/12.$$

Here, a, $b \in A_0$, $D = \frac{\partial}{\partial \tau} + \tau \frac{\partial}{\partial t}$, and $\rho : A_1/\partial A_0 \to \Lambda_0$ is the superresidue isomorphism of Λ_0-modules, changing parity (see the next section for its systematic theory), normalized by $\rho(t^{-1}\tau) = 1$.

The following map defines an isomorphism, functorial in Λ:

$$\mathrm{NS}_{\Lambda,0} \to K_\Lambda : ae_i \to at^{i+1}, \quad \lambda f_\alpha \to 2\lambda t^{\alpha+1/2}\tau, \quad z \to z,$$

where $a \in \Lambda_0$, $\lambda \in \Lambda_1$.

This means that the contact Lie algebra (in dimension $1|1$) is the universal even part of the Neveu–Schwarz superalgebras. We shall now show how to see K in E^s and therefore how to construct the (even part of) the Neveu-Schwarz sheaves on SUSY_1-families.

5.10. THE NEVEU–SCHWARZ EXTENSION. Using Lemma 5.6(b), we can construct an exact sequence

$$0 \to \omega_0 \to \overline{K_n} \to T_0 \to 0,$$

where T_0 is the even part of $T = \omega^{-2}$, and the third arrow is the symbol map $\sigma_{n/2} : G_{a,c} \to aD^2$. Again, the kernel of this map $\{G_{0,c} | c \in A_1\}$ can be identified with ω_0 via $\sigma_{(n-3)/2} : G_{0,c} \to dZ$. By putting $\delta a = dZ \cdot D_Z a$, $K_n = \overline{K_n}/\delta A_0$, we obtain an exact sequence

$$(5.5) \qquad 0 \to H_0^s \to K_n \to T_0 \to 0.$$

Now, a priori T_0 is not a Lie algebra. In order to see a natural bracket on it, consider the $1|1$-contact Lie algebra in its standard definition:

$$T' = \{X \in (AD + AD^2)_0 | [X, gD] \in A_0 D \text{ for all } g \in A_0\}.$$

Then

$$T' = \{bD^2 + Db \cdot D/2 | b \in A_0\}.$$

The symbol map $bD^2 + Db \cdot D/2 \to bD^2$ identifies T' and T_0. The induced bracket on T_0 is

$$[X_f, X_g] = X_{\{f,g\}}, \qquad \{f, g\} = Df \cdot Dg/2 + fD^2 g - gD^2 f,$$

which explains our formal definition of K_Λ.

Now we obtain our final result.

5.11. THEOREM. *The middle term in Eq. (5.5) is endowed with a natural structure of a Lie algebra, H_0^s lies in its center, and the symbol map is a homomorphism. This extension corresponds to the cocycle*

$$c(X_b, X_a) = \frac{(n-1)(n-3)}{16} aD^5 b \cdot dZ \mod \delta A_0.$$

Proof. As above, T' acts upon E^s via adjoint representation. From the definition of T' it follows that this action conserves the filtration. Hence, T' acts upon ω^i and $\overline{K}_n$. Concretely,

$$\mathrm{Lie}_{X_f}(g(dZ)^m) = (Df \cdot Dg/2 - fD^2 g + mgD^2 f/2)(dZ)^m.$$

Therefore,

$$\mathrm{Lie}_{X_b}(G_{a,c}) = G_{a',c'},$$

where $a' = \{b, a\}$ and b' is the coefficient at dZ in our cocycle.

The proof is now finished exactly as in the pure even case.

6. The Second Construction of the Neveu–Schwarz Sheaves

6.1. SHEAF ω AND DIVISORS ON A SUSY-CURVE. Let $\pi : X \to S$ be a SUSY$_1$-family. As in Section 5, we put $\omega = \omega_{X/S} = (\mathcal{T}^1_{X/S})^*$ and denote by $\delta : \mathcal{O}_X \to \omega$ the differential operator, which in any compatible coordinate system Z takes the form $\delta f = dZ \cdot D_Z f$. In more conceptual terms, it is the composition of the exterior differential $d_{X/S} : \mathcal{O}_X \to \Omega^1_{X/S}$ and the surjection $\Omega^1_{X/S} \to \omega$ dual to the inclusion $\mathcal{T}^1 \subset \mathcal{T}_{X/S}$ (cf. Section 2.1). Notice also that the structural isomorphism of $\mathcal{T}/\mathcal{T}^1$ with $(\mathcal{T}^1)^{\otimes 2}$ allows us to identify canonically $\omega_{X/S}$ with $\mathrm{Ber}(\Omega^1_{X/S})$.

Now let $\sigma : S \to X$ be an S-point of X, i.e., a section of π. The image of σ is a closed subspace of X of codimension $1|1$: Points of a SUSY-curve are not divisors. However, the SUSY-structure allows us to embed any point canonically into a divisor with the same reduction. Namely, let $Z = (z, \zeta)$ be a compatible coordinate and let $z - z_0 = \zeta - \zeta_0 = 0$ be the local equations of a point. Then the divisor $z - z_0 - \zeta\zeta_0 = 0$ does not depend on the choice of Z and hence is well-defined globally. We shall prove this in a universal setting, defining the relative "superdiagonal" $\Delta^s \subset X \times_S X$.

6.2. SUPERDIAGONAL. Denote by J the ideal in the structure sheaf of $X \times_S X$ defining the diagonal $i : \Delta \to X \to X \times_S X$. Let $\Delta^{(1)}$ be defined by J^2. We have the usual exact sequence

$$0 \to J/J^2 \to \mathcal{O}_{\Delta^{(1)}} \to \mathcal{O}_\Delta \to 0.$$

Furthermore, $J/J^2 = i_*(\Omega^1_{X/S})$. Put $I = \mathrm{Ker}(\Omega^1_{X/S} \to \omega_{X/S})$ and

$$\mathcal{O}_{\Delta^s} = \mathcal{O}_{\Delta^{(1)}}/i_*(I).$$

We shall often identify $\mathcal{O}_\Delta$-modules with sheaves on X.

6.3. LEMMA. *Δ^s is a closed analytic subspace of codimension $1|0$ called the (relative) superdiagonal of X. For a compatible coordinate $Z = (z,\zeta)$ put $Z_i = (z_i,\zeta_i) = p_i^*(Z)$, where p_i are two projections $X \times_S X \to X$. Then Δ^s is locally defined by the equation*

$$z_2 - z_1 - \zeta_2\zeta_1 = 0.$$

Proof. Under the standard identification we have

$$z_2 - z_1 - \zeta_2\zeta_1 \mod J^2 = dz - d\zeta \cdot \zeta.$$

(We recall that $d = d_{X/S}$). Moreover, the image of $dz - d\zeta\cdot\zeta$ in ω equals $D_Z z - D_Z\zeta\cdot\zeta = 0$. Hence, $(z_2 - z_1 - \zeta_2\zeta_1) \mod J^2 \in I \mod J^2$. But one easily checks locally that $I = (z_2 - z_1 - \zeta_2\zeta_1) + J^2$. It remains to verify that $J^2 \subset (z_2 - z_1 - \zeta_2\zeta_1)$. In fact, $J^2 = ((z_2 - z_1)^2, (z_2 - z_1)(\zeta_2 - \zeta_1))$, and we have

$$(z_2 - z_1)^2 = (z_2 - z_1 - \zeta_2\zeta_1)(z_2 - z_1 + \zeta_2\zeta_1);$$
$$(z_2 - z_1)(\zeta_2 - \zeta_1) = (z_2 - z_1 - \zeta_2\zeta_1)(\zeta_2 - \zeta_1).$$

Now we shall define the superresidue map ress. We start in a formal setting. Let A be a supercommutative ring of constants.

6.4. LEMMA. *Let $\delta : A((z,\zeta)) = 0 \to dZ \cdot A((a,\zeta)) = \omega$ be the map $\delta f = dZ \cdot D_Z f$. Define* ress : $\omega \to A$ *as a continuous A-linear map for which*

$$\mathrm{ress}(dZ \cdot z^a\zeta^b) = \begin{cases} 1 \text{ for } a = -1,\ b = 1; \\ 0 \text{ otherwise.} \end{cases}$$

Then we have an exact sequence

$$(6.1) \qquad 0 \to A \to 0 \xrightarrow{\delta} \omega \xrightarrow{\mathrm{ress}} A \to 0,$$

which does not depend on the choice of formal coordinates compatible with the same formal SUSY*-structure as Z.*

Proof. By definition, $\delta(z^a) = dZ \cdot az^{a-1}$, $\delta(z^b\zeta) = dZ \cdot Z^b$. It follows that Eq. (6.1) is exact. The independence of ress on Z will follow if we establish a formula of the type

$$dZ \cdot z^{-1}\zeta - dZ' \cdot z'^{-1}\zeta' = \delta L(Z, Z')$$

for two compatible SUSY-coordinates Z, Z'. By using the analytical identity $D_Z(\log z) = z^{-1}\zeta$ valid outside $z = 0$, one can guess and then easily prove a formal identity: If $z/z' \equiv 1 \mod (z, \zeta)$ then

$$dZ \cdot z^{-1}\zeta - dZ' \cdot z'^{-1}\zeta' = \delta \log(zz'^{-1}).$$

The general case reduces to this one by a linear coordinate change. In fact, put

$$\zeta' = (\sum_{i\geq 0} a_i z^i)\zeta + \sum_{j\geq 0} \alpha_j z^j = f(z)\zeta + \gamma(z);$$
$$z' = \sum_{k\geq 1} b_k z^k + (\sum_{l\geq 0} \beta_l z^l)\zeta = g(z) + \beta(z)\zeta.$$

Since Z and Z' define the same SUSY-structure, it follows from Lemma 4.3(b) that $D_z z' = \zeta' D_Z \zeta'$, i.e.,

$$\partial g/\partial z = f^2 \gamma \partial\gamma/\partial z; \qquad \beta = -f\gamma.$$

In particular, $b_1 = a_1^2 +$ nilpotent, so that we can define a new compatible system $Z'' = (b_1^{-1}z', b_1^{-1/2}\zeta')$. Replacing Z' by Z'' does not change the superresidue, and $z''z'^{-1} \equiv 1 \mod (z, \zeta)$.

6.5. A Local Calculation. Now let $\sigma : S \to X$ be an S-point locally defined by $z = z_0$, $\zeta = \zeta_0$. Consider a section ν of ω meromorphic in a neighborhood of this point and having a pole of order $\leq i + 1$ at the associated divisor, i.e.,

(6.2) $$\nu = dZ \cdot f(Z, s)(z - z_0 - \zeta\zeta_0)^{-(i+1)},$$

where f is regular. Denote by $\mathrm{ress}_\sigma(\nu)$ the superresidue calculated, say, in the completion $\pi^{-1}(\mathcal{O}_S)((z - z_0 - \zeta\zeta_0, \zeta - \zeta_0))$. Then we have

(6.3) $$\mathrm{ress}_\sigma(\nu) = \frac{1}{i!} D_Z^{2i+1}(f)|_{Z=Z_0}$$

In fact, if

$$f = \sum_{j \geq 0} a_j (z - z_0 - \zeta\zeta_0)^j + \sum_{k \geq 0} b_k (z - z_0 - \zeta\zeta_0)^k (\zeta - \zeta_0),$$

then, $\mathrm{ress}_\sigma(\nu) = (-1)^{\hat{b}_i} b_i$. On the other hand, $D_Z^{2i+1} = D_Z(\partial/\partial z)^i$ so that

$$D_Z^{2i+1}(f)|_{Z=Z_0} = (-1)^{\hat{b}_i} i! b_i.$$

6.6. Residue with Coefficients. The invariance property of the (super)residue shows that for any coherent sheaf E on S and any S-point of X, there exists a map of sheaves

$$\mathrm{ress}_\sigma : \omega \otimes \pi^* E(\infty D) \to E.$$

where D is the divisor associated to this S-point. Applying this to the families $p_{1,2} : X \times_S X \to X$, $D =$ superdiagonal and $E = \omega$, we get two residues:

$$\mathrm{ress}^{1,2} : \omega \boxtimes \omega(\infty\Delta^s) \to \mathcal{O}_X.$$

In fact, they coincide up to a total δ-differential.

6.7. Lemma. *There exists a unique map*

$$R : \omega \boxtimes \omega(\infty\Delta^s) \to \mathcal{O}_X.$$

such that $\delta \circ R = \mathrm{ress}^1 - \mathrm{ress}^2$ *and the restriction of* R *to sections with bounded order of pole is a differential operator along fibers of* $p_{1,2}$.

Proof. If there were two operators with such a property, their difference would be a differential operator mapping $\omega \boxtimes \omega(m\Delta^s)$ into the constants $\pi^{-1}(\mathcal{O}_S)$. But there are nonzero operators of this kind. Therefore, it suffices to construct R locally. Putting

$$\nu = dZ_1 \boxtimes dZ_2 \, f(Z_1, Z_2, s)(z_1 - z_2 - \zeta_1\zeta_2)^{-(i+1)}$$

and calculating locally via Eqs. (6.2) and (6.3), we find

$$\mathrm{ress}^1(\nu) = -dZ_2 \frac{1}{1!} D_{Z_1}^{2i+1} f(Z_1, Z_2, s)|_{Z_1 = Z_2 = Z},$$

$$\mathrm{ress}^2(\nu) = (-1)^{i+1} dZ_1 \frac{1}{i!} D_{Z_2}^{2i+1} f(Z_1, Z_2, s)|_{Z_1 = Z_2 = Z}.$$

Now put

$$R(\nu) = \frac{1}{i!}\sum_{b=0}^{2i}(-1)^{[(b-1)/2]}D_{Z_1}^{2i-b}D_{Z_2}^{b}\, f(Z_1, Z_2, s)|_{Z_1=Z_2=Z}.$$

In order to check that $\delta R(\nu) = (\mathrm{ress}^1 - \mathrm{ress}^2)(\nu)$, we write

$$f_{ab} = D_{Z_1}^{a} D_{Z_2}^{b} f\,|_{Z_1=Z_2=Z}.$$

Then

$$D_Z f_{ab} = f_{a+1,b} + (-1)^a f_{a,b+1}.$$

This leads to the cancellation of all terms in $\delta R(\nu)$ except for the first and the last.

6.8. THE GROTHENDIECK–SATO DESCRIPTION OF THE DIFFERENTIAL OPERATOR SHEAF. Let E be a locally free sheaf on X. Denote by $\mathcal{D}_{E/S}$ the sheaf of differential operators vertical over S and acting upon E on the left. The map defined in Section 5.4,

$$E \otimes \mathcal{D}_{\mathcal{O}_X/S} \otimes E^* \to \mathcal{D}_{E/S},$$

is an isomorphism. Put again $E^t = \omega \otimes E^*$ and consider the sheaf $E \boxtimes E^t(\infty\Delta^s)$ of meromorphic sections with poles on the superdiagonal. For a section ν of this sheaf, denote by $r(\nu)$ the operator $E \to E$ defined by

$$r(\nu)1 = \mathrm{ress}^2_{\Delta^s}(\nu, p_2^*(1)),$$

where $(\nu, p_2^*(1))$ denotes the result of the contraction $E \boxtimes (\omega \otimes E^* \otimes E) \to E \boxtimes \omega$.

6.9. LEMMA. *(a)* $\mathrm{Ker}(r) = E \boxtimes E^t$.
(b) r defines an isomorphism

$$E \boxtimes E^t((i+1)\Delta^s)/E \boxtimes E^t \to \mathcal{D}_{E/S}^{\leq(2i+1)},$$

where $\leq (2i+1)$ refers to the superorder of the differential operator (cf. Sections 4.2 and 4.3).

Proof. The first statement is clear and the second follows from Eq. (6.3). In fact, $\mathcal{D}_{E/S}$ is generated by sections of $\mathrm{End}\, E$ multiplied by powers of D_Z.

Odd powers are covered by r due to Eq. (6.3). Combining this with multiplication by $\zeta_1 - \zeta_2$ one also recovers even powers.

In this way, we get an extension of $\mathcal{D}_{E/S}$ of which this sheaf appears as a quotient, while $\mathcal{E}_{E/S}$ obtained by the method of Section 5 naturally contains $\mathcal{D}_{E/S}$.

In order to use it for a construction of the Neveu–Schwarz sheaves, we shall also need another version of adjointness (cf. Section 5.4).

Denote by $\overleftarrow{\mathcal{D}}_{E^t/S}$ the sheaf of differential operators over S acting upon E^t on the right.

6.10. LEMMA. *There exists a unique ring isomorphism*

$$D_{E/S} \to \overleftarrow{D}_{E^t/S} : P \to \overleftarrow{P}$$

and a unique map

$$\{\cdot, \cdot, \cdot\} : E^t \times D_{E/S} \times E \to \mathcal{O}_X$$

with the following properties: for every $F^t \in E^t$, $e \in E$, $P, Q \in \mathcal{D}_{E/S}$,

$$\langle f^t, Pe\rangle = \langle f^t \overleftarrow{P}, e\rangle + \delta\{f^t, P, e\}; \tag{6.5}$$

$$P \in \operatorname{End} E \Rightarrow \{f^t, P, e\} = 0; \tag{6.6}$$

$$\{f^t, D_Z, e\} = (-1)^{\hat{f}+1}(dZ)^{-1}\langle f^t, e\rangle; \tag{6.7}$$

$$\{f^t, QP, e\} = \{f^t \overleftarrow{Q}, P, e\} + \{f^t, Q, Pe\}. \tag{6.8}$$

SKETCH OF PROOF. From Eqs. (6.6)–(6.8), uniqueness of $\{f^t, P, e\}$ follows immediately. From Eq. (6.5) then follows uniqueness of $\overleftarrow{P}$. To prove the existence, one writes down an explicit formula in the local coordinates Z using the "integration by parts" procedure, and then checks all identities.

6.11. A CENTRAL EXTENSION OF $\mathcal{D}^{\mathrm{Lie}}_{E/S}$. Using Sections 6.9 and 6.10, we can now describe a canonical central extension of $\mathcal{D}_{E/S}$ considered as a *sheaf of Lie superalgebras*.

This is a more general construction than that presented in Section 5 in two respects: All differential operators are involved rather than only vector fields, and their coefficients now lie in a matrix sheaf $\operatorname{End} E$ rather than $\mathcal{O}_X$. Therefore, it is a simultaneous sheafification (on the curves of arbitrary genus) of the Virasoro (Neveu–Schwarz) and Kac–Moody Lie algebras. See [BS] for more details and local calculations in the pure even case.

We start from the exact sequence (see Lemma 6.9)

$$(6.9)\qquad 0 \to E \boxtimes E^t \to E \boxtimes E^t(\infty\Delta^s) \xrightarrow{r} \mathcal{D}_{E/S} \to 0.$$

Lemma 6.10 allows us to define an action of $\mathcal{D}^{\mathrm{Lie}}_{E/S}$ upon $E \boxtimes E^t(\infty\Delta^s)$ by

$$(6.10)\qquad \mathrm{Lie}(P)\nu = p_1^*(P)\nu - (-1)^{\hat{P}\hat{\nu}}\nu p_2^*(\overleftarrow{P})$$

for $P \in \mathcal{D}_{E/S}$, $\nu \in E \boxtimes E^t(\infty\Delta^s)$.

6.12. PROPOSITION. *(a) This action transforms $E \boxtimes E^t$ into itself and induces upon* $\mathrm{Im}(r) = \mathcal{D}_{E/S}$ *the adjoint action.*

(b) Let i be the embedding of the relative diagonal of X and j the composite map

$$E \boxtimes E^t \xrightarrow{i^*} E \otimes \omega \otimes E^* \xrightarrow{\mathrm{str}} \omega.$$

Then

$$\mathcal{D}^{\mathrm{Lie}}_{E/S}(E \boxtimes E^t) \subset j^{-1}(\delta\mathcal{O}_X).$$

Therefore, factorizing Eq. (6.9) *by $j^{-1}(\delta\mathcal{O}_X)$, we get a central extension of Lie superalgebras on X:*

$$0 \to H \to \overline{\mathcal{D}}^{\mathrm{Lie}}_{E/S} \to \mathcal{D}^{\mathrm{Lie}}_{E/S} \to 0, \qquad H = \omega_{X/S}/\delta\mathcal{O}_X.$$

Proof. (a) We must establish that for $e \in E$, we have

$$r(p_1^*(P)\nu)(e) - (-1)^{\hat{P}\hat{\nu}} r(\nu p_2^*(\overleftarrow{P}))(e) = Pr(\nu)(e) - (-1)^{\hat{P}\hat{\nu}} r(\nu)P(e).$$

The first members of both parts coincide. The second members also coincide in view of the adjunction formula and the fact that $\mathrm{ress} \circ \delta = 0$.

(b) Similarly, for $e \in E$, $f^t \in E^t$, we have

$$(6.11)\; i^*(Pe \boxtimes f^t - (-1)^{\hat{P}(\hat{e}+\hat{f})} e \boxtimes f^t\overleftarrow{P}) = (Pe) \otimes f^t - (-1)^{\hat{P}(\hat{e}+\hat{f})} e \otimes (f^t\overleftarrow{P}).$$

Furthermore,

$$\mathrm{str}(e \otimes f^t) = (-1)^{\hat{e}\hat{f}} \langle f^t, e \rangle$$

(one may take this for the definition of the supertrace). Therefore, after application of str, the r.h.s. of Eq. (6.11) becomes

$$(-1)^{\hat{f}(\hat{P}+\hat{e})}\langle f^t, Pe\rangle - (-1)^{\hat{f}(\hat{P}+\hat{e})}\langle f^t \overleftarrow{P}, e\rangle \in \delta\mathcal{O}_X.$$

7. Elliptic SUSY-Families

7.1. WHAT IS THE JACOBIAN OF A SUSY-CURVE? This is one of the major gaps in our understanding of this remarkable extension of the theory of Riemann surfaces to supergeometry. Of course, one can and must investigate an obvious definition of the (relative) jacobian via moduli space of invertible sheaves. However, this seems to be too simple-minded.

In the rest of this chapter, we shall describe some results of the theory of elliptic SUSY-curves that shed some light on the general problem and suggest some new questions. This material is taken from the Ph.D. thesis of A. Levin (Moscow University, 1988; cf. also his short notes in *Funkc. Analiz i ego Priloz.*).

Levin described two SUSY-families of genus 1, parametrized by the left half-plane and its superextension (they correspond to two possible parities of the structural theta-characteristic) and the period lattices, theta-functions, standard projective embedding, and so on. Before explaining his constructions in some detail, we shall summarize the most striking features that seem to be relevant generally.

(1) *An odd* SUSY-*curve of genus one (that is, a* SUSY-*curve with odd structural characteristic) is not an algebraic group.*

To be more precise, its universal covering is a group, but it is slightly noncommutative (the even part of its Lie superalgebra coincides with its center), and the period lattice *is not a normal subgroup*. Hence after factorization the group structure is lost. In all probability, an elliptic SUSY-curve should be its own jacobian. If this is so, the jacobians we are looking for need not be a priori group superschemes.

The genus one case suggests the following general definition of superanalytic tori. Consider the class of supergroup structures on $C^{m|n}$ whose Lie superalgebras are pseudoabelian, that is, with central even part. Then a general superanalytic torus is, by definition, a factor space of $C^{m|n}$ by, say, right shifts by elements of a discrete subgroup of rank 2m (of course, one should immediately pass to families, since interesting subgroups of this kind necessarily involve odd constants). In particular, they cannot parametrize invertible sheaves unless they are abelian groups.

One must now try to guess a right definition of superabelian varieties. Classically, they are precisely those tori that are embeddable into a projective space by means of suitable theta-functions. Here the genus one case teaches us the following lessons.

(2) *Levin's superthetafunctions of genus one are not pure even superfunctions: they have nontrivial even and odd components.*

Only in this way does one achieve a really nice generalization of classical formulas.

But these formulas then show that the superthetafunctions are not sections of an invertible sheaf but rather sections of a $1|1$-dimensional supervector bundle, endowed with an odd symmetry. I have suggested in [Ma1], Chapter 5, Section 6.4, that such bundles (or their section sheaves) can be considered as a substitute of invertible sheaves in supergeometry. An appropriate version of projective spaces carrying a structural sheaf "$\mathcal{O}(1)$" of this kind is given by Π-grassmannians $G\Pi(1|1;\mathbb{C}^{n|n},p) = \mathbb{P}_{\Pi}^{n-1}$ (cf. [Ma1], Chapter 5, Section 6.4, and Chapter 3 of this book).

Confirming the relevance of these objects for the theory of SUSY-curves, Levin proves that

(3) *Elliptic* SUSY-*curves can be naturally embedded into (products of) Π-projective spaces $\mathbb{P}_{\Pi}^{n}$.*

Keeping this in mind, we shall finish this chapter with a short summary of basic properties of Π-invertible sheaves due to Skornyakov.

7.2. CLASSICAL UNIFORMIZATION OF ELLIPTIC CURVES. We start with some classical constructions. By slightly changing the notation used in Section 1.14, we put $H = H^{1|0} = \{z \in \mathbb{C} | \operatorname{Re}(z) < 0\}$ and consider the following action of the group $\mathbb{Z}^2 = \mathbb{Z}A + \mathbb{Z}B$ upon $\mathbb{C} \times H$:

$$A : (z,t) \to (z + 2\pi i, t); \qquad B : (z,t) \to (z + t, t).$$

Putting $X = \mathbb{Z}^2 \setminus (\mathbb{C} \times H)$, we obtain a family of elliptic curves $\pi : X \to H$ with the following modular property: Any family with simply-connected base can be induced by π.

The sheaf $\Omega^1_{\mathbb{C}\times H/H}$ can be trivialized by the $\mathbb{Z}^2$-invariant section dz. A theta-characteristic can be described by an action of $\mathbb{Z}^2$ on the structure sheaf of $\mathbb{C} \times H$ differing from the standard action by signs that constitute a cocycle with values in $\mathbb{Z}/2\mathbb{Z}$. A classical notation for a theta-characteristic is $\begin{bmatrix} \delta \\ \varepsilon \end{bmatrix}$; $\delta, \varepsilon \in \{0, 1/2\}$. As a sheaf on X, it is the quotient of $\mathcal{O}_{\mathbb{C}\times H}$ with respect to the action

$$(Af)(z,t) = \exp(2\pi i(\delta - 1/2))f(z + 2\pi i, t);$$
$$(Bf)(z,t) = \exp(2\pi i(-\varepsilon + 1/2))f(z + t, t).$$

The parity of this theta-characteristic is $4\delta\varepsilon \mod 2$;

$$R^0\pi_*\begin{bmatrix}\delta\\ \varepsilon\end{bmatrix} = \begin{cases} 0 & \text{(even theta-char.)}\\ \mathcal{O}_H & \text{(odd theta-char.)};\end{cases}$$
$$R^1\pi_*\begin{bmatrix}\delta\\ \varepsilon\end{bmatrix} = \left(R^0\pi_*\begin{bmatrix}\delta\\ \varepsilon\end{bmatrix}\right)^* \quad \text{(Serre)}.$$

We shall now construct two families of elliptic SUSY-curves.

7.3. EVEN FAMILY. Consider $\mathbb{C}^{1|1}$ with the standard SUSY-structure $D_Z = \pi\zeta + \zeta\frac{\partial}{\partial z}$ endowed with the complex Lie supergroup law

$$(z,\zeta)\circ(z',\zeta') = (z+z'+\zeta\zeta', \zeta+\zeta').$$

The SUSY-structure is invariant with respect to:

(a) Right shifts.
(b) The group automorphism $c : (z,\zeta) \to (z,-\zeta)$.

Define now an action of $\mathbb{Z}^2 = \mathbb{Z}^2_{\text{ev}}$ upon $\mathbb{C}^{1|1} \times H$:

$$A_{\text{ev}} : (z,\zeta;t) \to (z+2\pi i,\zeta;t) = ((z,\zeta)\circ(2\pi i\cdot 0);t);$$
$$B_{\text{ev}} : (z,\zeta;t) \to (z+t,-\zeta;t) = (c(z,\zeta)\circ(t,0);t).$$

Since this action is discrete and compatible with the SUSY-structure, we get after factorization a SUSY-family

$$X_{\text{ev}} = \mathbb{C}^{1|1} \times H/\mathbb{Z}^2_{\text{ev}}.$$

Its theta-characteristic can be calculated by looking at the action of A and B on D_Z, the section that trivializes the sheaf dual to ω. We obtain $\begin{bmatrix}1/2\\ 0\end{bmatrix}$, which is even.

7.4. ODD FAMILY. This family is parametrized by $\mathcal{H} = H^{1|1} = H \times \mathbb{C}^{0|1}$ with coordinates $T = (t,\tau)$. The action of the period lattice upon $\mathbb{C}^{1|1} \times \mathcal{H}$ with the same relative SUSY-structure as in Section 7.3 is given by

$$A_{\text{odd}} : (z,\zeta;T) = (Z;T) \to (Z\circ(2\pi i,0);T) = (z+2\pi i,\zeta;t,\tau);$$
$$B_{\text{odd}} : (Z;T) \to (Z\circ T;T) = (z+t+\zeta\tau,\zeta+\tau;t,\tau).$$

As in Section 7.3, after factorization, we obtain an odd SUSY-family $X_{\text{odd}} \to \mathcal{H}$ with the structural theta-characteristic $\begin{bmatrix}1/2\\ 1/2\end{bmatrix}$.

7.5. THEOREM. *Any* SUSY-*family of genus one with a connected simply-connected base can be induced from either* X_{ev} *or* X_{odd}.

Proof. A SUSY-family as in the statement will be called a local family. Consider a local family Y/W. The reduced family admits a global section. By using it as a fiberwise initial point, one can construct a fiberwise universal covering space $\overline{Y}/W$, which is a SUSY-family of $\mathbb{C}^{1|1}$'s over W upon which $\mathbb{Z}^2$ acts discretely.

In order to prove the theorem, it suffices to construct on $\overline{Y}/W$ a relative global coordinate $Z = (z, \zeta)$ compatible with the SUSY-structure and an even function t on W (for an even family) or a pair of functions $T = (t, \tau)$ (for an odd family) such that the action of $\mathbb{Z}^2$ on $\overline{Y}/W$ could be described by the formulas of Section 7.3 or 7.4.

Assume first that W is pure even. Then from Proposition 2.3, it follows that a global compatible coordinate system Z on $\overline{Y}/W$ does exist. Writing A_1, A_2 instead of A, B (the generators of $\mathbb{Z}^2$), we have $A_i^*(z) = f_i(z)$, $A_i^*(\zeta) = \zeta g_i(z)$, since there are no odd constants. Moreover, $f_i(z) = z + t_i$, where t_i are functions on W. Compatibility of A_i with the SUSY-structure immediately shows that $g_i = \pm 1$. Let us consider the possible four cases separately.

(i) $g_1 = g_2 = 1$. If $\mathrm{Re}(2\pi i t_2/t_1)_{\mathrm{red}} < 0$, replace z by $2\pi i z/t_1$, ζ by $(2\pi i/t_1)^{1/2}\zeta$ and put $t = 2\pi i t_2/t_1$, $\tau = 0$. This defines the desired morphism to X_{odd}. If $\mathrm{Re}(2\pi i t_2/t_1) > 0$, replace first A_2 by $-A_2$.

(ii) $g_1 = -g_2 = 1$. The same replacements as in (i) define a morphism to X_{even}.

(iii) Reduces to (ii) by $A_1 \to A_2$, $A_2 \to A_1$.

(iv) Reduces to (i) by replacing A_1 by $A_1 + A_2$.

We shall use the case of a pure even base W thus treated as the induction base. Generally, denote the sheaf of nilpotents in $\mathcal{O}_W$ by N and consecutively construct our coordinates modulo the powers of p^*N, where $p : \overline{Y} \to W$ is the structural projection. The inductive steps starting from $(p^*N)^i$ differ slightly depending on the parity of i. For brevity, we shall give details only for even families.

7.4.a. Even Family, Odd Step. Assume that we already have on $\overline{Y} \to W$ a compatible system (z, ζ) that has the necessary properties modulo $(p^*N)^{2i-1}$. Let

$$A^*(z,\zeta) \equiv (z + 2\pi i + \eta\psi_A(z),\ \eta + \varphi_A(z)) \mod p^*N^{2i};$$
$$B^*(z,\zeta) \equiv (z + t + \eta\psi_B(z),\ -\eta - \varphi_b(v)) \mod p^*N^{2i}.$$

Here ψ_A, ψ_B, φ_A, φ_B are odd sections of $p^*N^{2i-1} \subset \mathcal{O}_{\overline{Y}}$. The compatibility with the SUSY-structure implies $\psi_A \equiv \varphi_A \mod p^*N^{2i}$ and similarly for ψ_B, φ_B. Since A and B commute, we have

$$\psi_B(z) - \psi_A(z+t) \equiv \psi_A(z) + \psi_B(z+2\pi i) \mod p^*N^{2i}.$$

This means that (ψ_A, ψ_B) define a group cocycle on $\mathbb{Z}^2$ with values in the $\mathbb{Z}^2$-module $\begin{bmatrix} 1/2 \\ 0 \end{bmatrix} \otimes_W p^*(N^{2i-1}/N^{2i})$. By interpreting it topologically and by using the Leray spectral sequence over the simply-connected base, one sees that this cocycle splits. Hence, we can find a function $\gamma \in p^*N^{2i-1}$ such that

$$\psi_A(z) \equiv \gamma(z+2\pi i) - \gamma(z) \mod p^*N^{2i};$$
$$\psi_B(z) \equiv -\gamma(z+t) - \gamma(z) \mod p^*N^{2i}.$$

If we now replace z by $z - \zeta\gamma(z)$, and ζ by $\zeta - \gamma(z)$, we shall get the necessary form of action of A, B modulo p^*N^{2i}. One checks directly that it is SUSY-compatible.

7.4.b. Even Family, Even Step. If A and B now act correctly modulo p^*N^{2i}, we may put

$$A^*(z\zeta) = (z + 2\pi i + \psi_A(z),\ \eta + \eta\varphi_A(z)) \mod p^*N^{2i+1};$$
$$B^*(z,\zeta) = (z + t + \psi_B(z),\ -\eta - \eta\varphi_B(z)) \mod p^*N^{2i+1}.$$

Here φ, ψ are even sections of p^*N^{2i}. From the compatibility with the SUSY-structure, one finds

$$2\varphi_A \equiv \frac{\partial\psi_A}{\partial z} \mod p^*N^{2i+1}; \qquad 2\varphi_B \equiv \frac{\partial\psi_B}{\partial z} \mod p^*N^{2i+1}.$$

Furthermore, A and B commute, so that

$$\psi_A(z) + \psi_B(z+2\pi i) \equiv \psi_B(z) + \psi_A(z+t) \mod p^*N^{2i+1}.$$

Hence, we again get a $\mathbb{Z}^2$-cocycle. Since the relevant theta-characteristic is now odd, it belongs to a class of form $(0, q)$, $q \in \Gamma(\mathcal{O}_W)$. This means that we can find function $\gamma \in p^*N^{2i}$ and $q \in N^{2i}$ such that

$$\varphi_A(z) \equiv \gamma(z+2\pi i) - \gamma(z) \mod p^*N^{2i+1};$$
$$\varphi_B(z) \equiv \gamma(z+t) - \gamma(z) + q \mod p^*N^{2i+1}.$$

Now replace z by $z - \zeta\gamma$, ζ by $\zeta(1 - \frac{\partial\gamma}{\partial z})^{1/2}$, and t by $t - q$. A direct calculation shows that A and B act properly in these coordinates modulo p^*N^{2i+1}.

Odd families are treated similarly.

8. Supertheta-Functions

8.1. SUPEREXPONENTIAL FUNCTION. In Section 7.3, we have defined a complex Lie supergroup structure on $\mathbb{C}^{1|1}$. Let us denote it by $\mathbb{G}_a^{1|1}$. It is a noncommutative version of the usual additive group $\mathbb{G}_a$. There is also a natural version of $\mathbb{G}_m$ that will be denoted by $\mathbb{G}_m^{1|1}$. As an algebraic supergroup, it represents the functor of points $\mathbb{G}_m^{1|1}(A) = A^*$ (invertible elements of a supercommutative ring A), while $\mathbb{G}_m(A) = A_0^*$. A natural coordinate system on $\mathbb{G}_m^{1|1}$ is $V = (v, \nu)$ corresponding to even and odd parts of a point as an element of A^*. The multiplication law is

$$(v, \nu)(v', \nu') = (vv' + \nu\nu', \; v\nu' + v'\nu).$$

A right-invariant SUSY-structure on $\mathbb{G}_m^{1|1}$ is given by

$$\mathcal{D}_V = v\frac{\partial}{\partial\eta} + \eta\frac{\partial}{\partial v}.$$

The superexponential is a complex-analytic morphism of supermanifolds $\text{Exp} : \mathbb{G}_a^{1|1} \to \mathbb{G}_m^{1|1}$, which on generic points is given by

$$Z = (z, \zeta) \to \text{Exp}(Z) = \exp(z)(1 + \zeta).$$

8.2. LEMMA. Exp *is a surjective morphism of groups compatible with* SUSY-*structures, with kernel* $(2\pi i\mathbb{Z}, 0)$.

The proof is straightforward. In particular,

$$d\,\text{Exp}(D_Z) = \mathcal{D}_{\text{Exp}(Z)}.$$

Using Exp, one can represent X_{ev}/H (resp. $X_{\text{odd}}/\mathcal{H}$) as a quotient space of $\mathbb{G}_m^{1|1} \times H$ (resp. $\mathbb{G}_m^{1|1} \times \mathcal{H}$) with respect to an action of $\mathbb{Z}$ instead of $\mathbb{Z} \times \mathbb{Z}$. The generator acts as follows:

$$\text{even: } (V, t) \to (c(V)\exp(t), t);$$
$$\text{odd: } (V, T) \to (V\,\text{Exp}(T), T).$$

Putting $q = \exp(t)$, $Q = \text{Exp}(T)$ and restricting the base space to $0 < |q| < 1$ (resp. $0 < |Q_{\text{red}}| < 1$), we get an analog of the Jacobi uniformization.

8.3. THETA-FUNCTIONS OF THE EVEN FAMILY. We put

$$\theta_\Pi^{\text{ev}}\begin{bmatrix}\delta\\ \varepsilon\end{bmatrix}(z, \zeta \mid t) = \sum_{m=-\infty}^{\infty} \text{Exp}\{(m+\delta)^2 t/2 + (m+\delta)(z + 2\pi i\varepsilon), \; (-1)^{m+\delta}\zeta\}.$$

This series converges uniformly on superdomains of $G_a^{1|1} \times H$ of the form $|\mathrm{Re}(z)| < c$, $\mathrm{Re}(t) < \varepsilon < 0$. In fact, we have

$$\theta_\Pi^{\mathrm{ev}}\begin{bmatrix}\delta\\ \varepsilon\end{bmatrix}(z,\zeta \mid t) = \theta\begin{bmatrix}\delta\\ \varepsilon\end{bmatrix}(z \mid t) + \zeta\theta\begin{bmatrix}\delta\\ \varepsilon+1/2\end{bmatrix}(z \mid t),$$

where the right-hand side is a sum of the classical theta-functions.

A straightforward calculation furnishes the following functional equations:

$$\begin{aligned}\theta_\Pi^{\mathrm{ev}}&\begin{bmatrix}\delta\\ e\end{bmatrix}(z+mt+2\pi i n,\ (-1)^m\zeta \mid t)\\ &= \theta_\Pi^{\mathrm{ev}}\begin{bmatrix}\delta\\ \varepsilon\end{bmatrix}(z,\zeta \mid t)\exp(-m^2t/2 - mz - 2\pi i(m\varepsilon - n\delta)).\end{aligned}$$

$$\begin{aligned}\theta_\Pi^{\mathrm{ev}}&\begin{bmatrix}\delta\\ \varepsilon\end{bmatrix}(z,\zeta \mid t)\\ &= \theta_\Pi^{\mathrm{ev}}\begin{bmatrix}0\\ 0\end{bmatrix}(z+\delta t+2\pi i\varepsilon,\ (-1)^\delta\zeta \mid t)\exp(\delta^2 t/2 + \delta(z+2\pi i\varepsilon)).\end{aligned}$$

8.4. Theta-Functions of the Odd Family. Following Levin, we shall introduce into the definition an auxiliary even function A on $\mathcal{H}$. We put

$$\begin{aligned}\theta_\Pi^{\mathrm{odd}}\begin{bmatrix}\delta\\ \varepsilon\end{bmatrix}(z,\zeta \mid t,\tau) = \sum_{m=-\infty}^{\infty} \mathrm{Exp}\,\{&(m+\delta)t/2\\ &+ (m+\delta)(z+2\pi i\varepsilon) + \zeta\tau[A^2(m+\delta)/6\\ &+ (m+\delta)^2/2 - A^2(m+\delta)^3/6],\\ &A(m+\delta)^2\tau/2 + A(m+\delta)\zeta\}.\end{aligned}$$

Denoting by $'$ the z-derivative, we have the following decomposition into a linear combination of the classical thetas, showing in particular the usual convergence properties:

$$\begin{aligned}\theta_\Pi^{\mathrm{odd}}\begin{bmatrix}\delta\\ \varepsilon\end{bmatrix}(z,\zeta \mid t,\tau) =&\,\theta\begin{bmatrix}\delta\\ \varepsilon\end{bmatrix}(z \mid t) + \zeta\tau\{A^2/6\cdot\tau'\begin{bmatrix}\delta\\ \varepsilon\end{bmatrix}(z \mid t)\\ &+ 1/2\cdot\tau''\begin{bmatrix}\delta\\ \varepsilon\end{bmatrix}(z \mid t) - A^2/6\cdot\tau''\begin{bmatrix}\delta\\ \varepsilon\end{bmatrix}(z \mid t)\}\\ &+ A\zeta/2\cdot\tau''\begin{bmatrix}\delta\\ \varepsilon\end{bmatrix}(z \mid t) + A\zeta\tau'\begin{bmatrix}\delta\\ \varepsilon\end{bmatrix}(z \mid t).\end{aligned}$$

The most essential functional equation now shows that a shift by a nonconstant period multiplies θ^{odd} on the right by a multiplier with nontrivial even and odd components:

$$\theta_\Pi^{\text{odd}}\begin{bmatrix}\delta\\ \varepsilon\end{bmatrix}(z+t+\zeta\tau,\ \zeta+\tau \mid t,\tau)$$
$$=\theta_\Pi^{\text{odd}}\begin{bmatrix}\delta\\ \varepsilon\end{bmatrix}(z,\zeta \mid t,\tau)\,\mathrm{Exp}(-t/2-z-\zeta\tau/2-2\pi i\varepsilon, -A\tau/2-A\zeta).$$

Shift by $2\pi i$ multiplies θ^{odd} by $\exp(2\pi i\delta)$. Finally

$$\theta_\Pi^{\text{odd}}\begin{bmatrix}\delta\\ \varepsilon\end{bmatrix}(z,\zeta \mid t,\tau)=\theta_\Pi^{\text{odd}}\begin{bmatrix}0\\ 0\end{bmatrix}(z+\delta t+2\pi i\varepsilon+\delta\zeta\tau,\ \zeta+\delta\tau \mid t,\tau)$$
$$\times\,\mathrm{Exp}\{\delta^2 t/2+\delta(z+2\pi i\varepsilon)$$
$$+\zeta\tau(A^2\delta/6+\delta^2/2-A^2\delta^3/6,\ A(\delta^2\tau/2+\delta\zeta))\}.$$

We omit the verifications.

8.5. Π-PROJECTIVE SPACE. Let $T=\mathbb{C}^{n|n}$ be a vector space endowed with an odd linear involution $p: T\to T$, $p^2=\mathrm{id}$. Denote by $P_\Pi^{n-1|n-1}=P_\Pi(T,p)=G\Pi(1|1;T,p)$ the grassmannian of p-invariant $1|1$-subspaces in T. Its tautological sheaf $\mathcal{S}$ is also endowed with an odd involution p.

Put $\mathcal{S}^*=\mathcal{O}_\Pi(1)$. Each element of T^* defines a section of $\mathcal{O}_\Pi(1)$. Let $(x_1,\dots,x_n,\xi_1,\dots,\xi_n)$ be a basis of T^* such that $p^*(x_i)=\xi_i$. Then $u_i=x_i+\xi_i$ constitute n p-invariant sections of $\mathcal{O}_\Pi(1)$ having no common zeros on $\mathbb{P}_{\Pi,\text{red}}$. Let M be a superspace, and $f: M\to\mathbb{P}_\Pi^{n-1|n-1}$ a morphism. Then $f^*(\mathcal{O}_\Pi(1))$ is a locally free sheaf of rank $1|1$ on M endowed with an odd involution $f^*(p)$ and n $f^*(p)$-invariant sections without common zeroes $f^*(u_i)$. Using these data, one reconstructs f uniquely. If one multiplies all sections $f^*(u_i)$ on the right simultaneously by the same invertible section of $\mathcal{O}_M^*$, *the reconstructed f does not change.*

Instead of $f^*(u_i)$, one can use n even sections $f^*(x_i)$; odd parts then can be regained with the help of the involution. We shall call a $1|1$-sheaf with an odd involution a Π-*invertible sheaf*. Giving an odd involution on a $1|1$-sheaf is equivalent to reducing its structure group from $\mathrm{GL}(1|1)$ to $\mathbb{G}_m^{1|1}$.

8.6. Π-INVERTIBLE SHEAVES ON X_{ev} AND X_{odd}. A Π-invertible sheaf on X_{ev} (resp. X_{odd}) can be defined by a twisted action of $\mathbb{Z}^2$ on the structure sheaf of $\mathbb{C}^{1|1}\times H$ (resp. $\mathbb{C}^{1|1}\times\mathcal{H}$). Twisting is given by a right multiplication by a not necessarily homogeneous function.

The functional equations for θ_Π with characteristics stated in Sections 8.3 and 8.4 determine simultaneously such sheaves (twisting is given by multipliers) and their sections (theta-function themselves). We shall now investigate the maps into Π-projective spaces defined by these theta-functions.

8.7. JACOBI FUNCTIONS FOR X_{ev}. For brevity, let us write $\theta^{\text{ev}}_{2\delta,2\varepsilon}$ instead of $\theta^{\text{ev}}_{\Pi}\begin{bmatrix}\delta\\ \varepsilon\end{bmatrix}$, and similarly for θ^{odd}. The functions $\theta^{\text{ev}}_{00}(2z,\zeta|2t)$ and $\theta^{\text{ev}}_{10}(2z,\zeta|2t)$ are sections of the Π-invertible sheaf determined by the multiplicators $A_{\text{ev}}\to 1$, $B_{\text{ev}}\to\exp(-t-2z)$. Hence, they define a map of X_{ev} to $\mathbb{P}^{1|1}\times H$.

In the same manner, another map of this kind is defined by $\theta^{\text{ev}}_{01}(2z,\zeta|2t)$ and $\theta^{\text{ev}}_{11}(2z,\zeta|2t)$. The following "Nullwerte" are classical elliptic functions on H:

$$k^{1/2}_{\text{ev}}(2t)=\theta^{\text{ev}}_{10}(0,0\mid 2t)/\theta^{\text{ev}}_{00}(0,0\mid 2t);$$
$$k'^{1/2}_{\text{ev}}(2t)=\theta^{\text{ev}}_{01}(0,0\mid 2t)/\theta^{\text{ev}}_{00}(0,0\mid 2t).$$

Using them for normalization sake, we also introduce the superversion of the Jacobi functions on $\mathbb{C}^{1|1}\times H$:

$$\mathrm{Sn}^{\text{ev}}(2z,\zeta\mid 2t)=k^{-1/2}_{\text{ev}}(2t)\theta^{\text{ev}}_{11}(2z,\zeta\mid 2t)/\theta^{\text{ev}}_{01}(2z,\zeta\mid 2t);$$
$$\mathrm{Cd}^{\text{ev}}(2a,\zeta\mid 2t)=k^{-1/2}_{\text{ev}}(2t)\theta^{\text{ev}}_{10}(2z,\zeta\mid 2t)/\theta^{\text{ev}}_{00}(2z,\zeta\mid 2t).$$

They should be considered as coordinate descriptions of two maps $X_{\text{ev}}\to\mathbb{P}^{1|1}_{\Pi}\times H$ in an affine chart.

Consider a relatively open subset in $\mathbb{P}^{1|1}\times\mathbb{P}^{1|1}\times H$ with coordinates $(v,\nu;v',\nu';t)$ corresponding to this chart.

8.8. THEOREM. *The map $X_{\text{ev}}\to\mathbb{P}^{1|1}\times\mathbb{P}^{1|1}\times H$ described above is a closed immersion of supermanifolds, and its image in this chart is given by the equations*

$$(k^2_{\text{ev}}v^2-1)(k^2_{\text{ev}}v'^2-1)=k'^2;$$
$$v(k'_{\text{ev}}\nu+(k^2_{\text{ev}}v^2-1)\nu')=v'(k'_{\text{ev}}\nu-(k^2_{\text{ev}}v^2-1)\nu').$$

Proof. In order to check these relations, we use the classical equations

$$k^2+k'^2=1;\qquad k^2sn^2+dn^2=1;\qquad k^2cn^2-dn^2=k'^2$$

and the component decompositions that follow from Section 8.3:

$$\begin{aligned}\mathrm{Sn}^{\text{ev}}(2z,\zeta\mid 2t)&=k^{-1/2}(2t)(\theta_{11}(2z\mid 2t)-\zeta\theta_{10}(2z\mid 2t))\\&\quad\times(\theta_{01}(2z\mid 2\theta_{00}(2z\mid 2t))^{-1}\\&=\mathrm{sn}(2z\mid 2t)-\zeta k'^{-1/2}(2t)\{\mathrm{cn}+\mathrm{sn}\,dn\}(2z\mid 2t);\end{aligned}$$

$$\mathrm{Cd}^{\text{ev}}(2z,\zeta\mid 2t)=\mathrm{cn}/dn(2z\mid 2t)+\zeta k'^2_{\text{ev}}(2t)\{\mathrm{sn}/dn-\mathrm{cn}/dn^2\}(2z\mid 2t).$$

It is known classically that the reduced map is a closed embedding. Hence, the Jacobian criterion reduces to checking that $\frac{\partial \mathrm{Sn}^{\mathrm{ev}}}{\partial \zeta} = k'^{1/2}(-\mathrm{cn} - \mathrm{sn}\,\mathrm{dn})$ and $\frac{\partial \mathrm{Cd}^{\mathrm{ev}}}{\partial \zeta} = k'(\mathrm{sn}/dn - \mathrm{cn}/dn^2)$ nowhere vanish simultaneously. In fact, cn and $\mathrm{sn}\,dn$ have no common zeroes.

It remains to check that the equations given in the statement define a $2|1$-dimensional supermanifold. Its reduced space is a 2-manifold. Hence one must only check that the second equation for the odd coordinates does not degenerate anywhere, that is, $k'(v - v')$ and $(v + v')(k^2v^2 - 1)$ do not vanish simultaneously. This follows from the invertibility of $k^2v^2 - 1$ (the first equation) and the fact that v and v' do not vanish simultaneously.

8.9. Jacobi Functions for X_{odd}. Four theta-functions

$$\theta^{\mathrm{odd}}_{2\delta,2\varepsilon}(2z, 2^{1/2}\zeta | t, 2^{-1/2}\tau)$$

can be considered as sections of a Π-invertible sheaf on X_{odd} determined by the multipliers $A_{\mathrm{odd}} \to 1$, $B_{\mathrm{odd}} \to \mathrm{Exp}(-2t - 2u - 2\zeta\tau + A^2\zeta\tau, -A(2^{1/2}\zeta + 2^{3/2}\zeta))$. Thus, they define a map $X_{\mathrm{odd}} \to \mathbb{P}^{3|3}_{\Pi} \times \mathcal{H}$.

Put

$$k^{1/2}_{\mathrm{odd}}(t, 2^{-1/2}\tau) = \theta^{\mathrm{odd}}_{10}(0, 0 \mid t, 2^{-1/2}\tau)/\theta^{\mathrm{odd}}_{00}(0, 0 \mid t, 2^{-1/2}\tau);$$
$$k'^{1/2}_{\mathrm{odd}}(t, 2^{-1/2}\tau) = \theta^{\mathrm{odd}}_{01}(0, 0 \mid t, 2^{-1/2}\tau)/\theta^{\mathrm{odd}}_{00}(0, 0 \mid t, 2^{-1/2}\tau);$$

From Section 8.4, we can obtain the following component decomposition

$$k^{1/2}_{\mathrm{odd}}(t, 2^{-1/2}\tau) = k^{1/2}(t) + 2^{-1/2}\tau A(t)\frac{\partial k^{1/2}}{\partial t};$$
$$k'^{1/2}_{\mathrm{odd}}(t, 2^{-1/2}\tau) = k'^{1/2}(t) + 2^{-1/2}\tau A(t)\frac{\partial k'^{1/2}}{\partial t}.$$

It follows that

$$k^2_{\mathrm{odd}} + k'^2_{\mathrm{odd}} = 1.$$

We shall now define three functions on $\mathbb{C}^{1|1} \times \mathcal{H}$ that can be viewed as a coordinate description of the map $X_{\mathrm{odd}} \to \mathbb{P}^{3|3} \times \mathcal{H}$:

$$\begin{aligned}\mathrm{Sn}^{\mathrm{odd}}&(2z, 2^{1/2}\zeta \mid t, 2^{-1/2}\tau)\\ &= k^{-1/2}_{\mathrm{odd}}(t, 2^{-1/2}\tau)(\theta^{\mathrm{odd}}_{11}/\theta^{\mathrm{odd}}_{01})(2z, 2^{1/2}\zeta \mid t, 2^{-1/2}\tau),\end{aligned}$$

$$\mathrm{Cn}^{\mathrm{odd}} = k^{-1/2}_{\mathrm{odd}} k'^{1/2}_{\mathrm{odd}} \theta^{\mathrm{odd}}_{10}/\theta^{\mathrm{odd}}_{01};$$
$$\mathrm{Dn}^{\mathrm{odd}} = k'^{1/2}_{\mathrm{odd}} \theta^{\mathrm{odd}}_{00}/\theta^{\mathrm{odd}}_{01}.$$

8.10. THEOREM. *The map* $X_{\text{odd}} \to \mathbb{P}^{3|3} \times \mathcal{H}$ *is a closed immersion of supermanifolds. Its image is defined by the following relations between* Sn^{odd}, Cn^{odd}, Dn^{odd}. *Put*

$$\mathcal{H} = k^{1/2}_{\text{odd},1}/k^{1/2}_{\text{odd},0}, \qquad \mathcal{H}' = k'^{1/2}_{\text{odd},1}/k'^{1/2}_{\text{odd},0}.$$

Then

$$\begin{aligned}
&\text{Dn}(\text{Sn}^2 + \text{Cn}^2 - 1) = ((\mathcal{H}' - \mathcal{H})\,\text{Cn}_0\,\text{Sn}_1 + \mathcal{H}\,\text{Sn}_0\,\text{Cn}_1)F;\\
&\text{Cn}((k\,\text{Sn})^2 + \text{Dn}^2 - 1) = (\mathcal{H}'\,\text{Dn}_0\,\text{Sn}_1 - \mathcal{H}\,\text{Sn}_0\,\text{Dn}_1)F;\\
&\text{Sn}((k'^{-1}\,\text{Dn})^2 - (k'^{-1}k\,\text{Cn})^2 - 1) = k'^{-2}(\mathcal{H}'\,\text{Dn}_0\,\text{Cn}_1\\
&\qquad\qquad + (\mathcal{H} - \mathcal{H}')\,\text{Cn}_0\,\text{Dn}_1)F;\\
&F = A^2(\pi\theta_{00}^2(0 \mid t)^{-1}).
\end{aligned}$$

For brevity, we omitted the sub- and superscripts odd; *the implied arguments are as in the definition.*

SKETCH OF PROOF. We start by giving the components of Sn^{odd}, Cn^{odd}, Dn^{odd}. They can be derived from Section 8.4 by using the following classical identities:

$$\begin{aligned}
&2\partial_t\tau = \tau'';\\
&\tau_{11}'''\theta_{01} - 3\tau_{11}''\tau_{01}' + 3\tau_{11}'\tau_{01}'' - \theta_{11}\tau'''_{01}\\
&\quad = \theta_{01}^2[(\theta_{11}\theta_{01}^{-1})''' - 6(\theta_{11}\theta_{01}^{-1})'(\log\theta_{01})'']
\end{aligned}$$

and also identities between sn, cn, dn, and their derivatives. As earlier, we put $\text{sn} = \text{sn}(2z|t) = k^{-1/2}(t)(\theta_{11}\theta_{01}^{-1})(2z|t)$. Its argument differs from the classical normalization by $K = 2^{-1}\pi\theta_{00}^2(0|t)$. We have

$$\begin{aligned}
\text{Sn}^{\text{odd}} &= \text{sn} + [A^2(\text{sn}' - \text{sn}''' + 6\,\text{sn}'(\log\theta_{01})'')/6 + k^{-1/2}\partial_t(\theta_{11}\theta_{01}^{-1})\\
&\quad - A^2\partial_t(k^{-1/2})(\theta_{11}\theta_{01}^{-1})']\zeta\tau + 2^{-1/2}A\tau\partial_t\,\text{sn} + 2^{1/2}A\tau\,\text{sn}'.\\
\text{Cn}^{\text{odd}} &= \text{cn} + [A^2(\text{cn}' - \text{cn}''' + 6\,\text{cn}'(\log\theta_{01})'')/6 + k'^{1/2}k^{-1/2}\partial_t(\theta_{10}\theta_{01}^{-1})\\
&\quad - A^2\partial_t(k'^{1/2}k^{-1/2})(\theta_{10}\theta_{01}^{-1})']\zeta\tau + 2^{-1/2}A\tau\partial_t\,\text{cn} + 2^{1/2}A\tau\,\text{cn}'.\\
\text{Dn}^{\text{odd}} &= \text{dn} + [A^2(\text{dn}' - \text{dn}''' + 6\,\text{dn}'(\log\theta_{01})'')/6 + k'^{1/2}\partial_t(\theta_{00}\theta_{01}^{-1})\\
&\quad - A^2\partial_t(k'^{1/2})(\theta_{10}\theta_{01}^{-1})']\zeta\tau + 2^{-1/2}A\tau\partial_t\,\text{dn} + 2^{1/2}A\tau\,\text{dn}'.
\end{aligned}$$

A calculation shows that

$$(\text{Sn}^{\text{odd}})^2 + (\text{Cn}^{\text{odd}})^2 = 1 + (\text{sn}''\,\text{sn} + \text{cn}''\,\text{cn})\zeta\tau.$$

From the classical identities, it follows that

$$\mathrm{sn}'' \,\mathrm{sn} + \mathrm{cn}'' \,\mathrm{cn} = -(\pi\theta_{00}^2(0 \mid t))^2 \,\mathrm{dn}^2 .$$

Moreover,

$$(\mathcal{H} - \mathcal{H}')\,\mathrm{Cn}_0^{\mathrm{odd}}\,\mathrm{Sn}_1^{\mathrm{odd}} + \mathcal{H}\,\mathrm{Sn}_0^{\mathrm{odd}}\,\mathrm{Cn}_1^{\mathrm{odd}} = -(\pi\theta_{00}^2(0 \mid t))^3 A^2\,\mathrm{dn}^3\,\zeta\tau.$$

From this, we find the first equation of the theorem. The other two are proved similarly.

The reduced equations define a family of elliptic curves over $\mathcal{H}_{\mathrm{red}}$. In order to check that the equations themselves define a supermanifold of dimension $2|2$, it suffices to verify that the matrix consisting of derivatives of the left-hand sides with respect to the odd variables $\mathrm{Sn}_1^{\mathrm{odd}}$, $\mathrm{Cn}_1^{\mathrm{odd}}$, $\mathrm{Dn}_1^{\mathrm{odd}}$ has rank 2 everywhere. But the reduced matrix is of the form

$$\begin{pmatrix} \mathrm{sn} & \mathrm{cn} & 0 \\ k^2\,\mathrm{sn} & 0 & \mathrm{dn} \\ 0 & -k^2k'^{-2}\,\mathrm{cn} & k'^{-2}\,\mathrm{dn} \end{pmatrix}$$

The columns are linearly dependent:

$$\mathrm{cn}\,\mathrm{dn}(I) - \mathrm{sn}\,\mathrm{dn}(II) - k^2\,\mathrm{sn}\,\mathrm{cn}(III) = 0.$$

On the other hand, since the sets of zeroes of sn, cn, dn are pairwise disjoint, at any point at least one of the 2×2-minors does not vanish.

Finally, the jacobian of the map $X_{\mathrm{odd}} \to \mathrm{P}^{3|3} \times \mathcal{H}$ that we are considering is of maximal rank everywhere, because

$$\partial_\zeta(\mathrm{Sn}_1^{\mathrm{odd}},\ \mathrm{Cn}_1^{\mathrm{odd}},\ \mathrm{Dn}_1^{\mathrm{odd}}) = 2^{1/2}A(\mathrm{sn}', \mathrm{cn}', \mathrm{dn}')$$

do not vanish simultaneously.

Let us notice that all formulas acquire the most classical form if one puts $A = \pi^{1/2}\theta_{00}(0|t)$. However, bookkeeping becomes slightly more transparent if this choice is postponed.

8.11. Π-INVERTIBLE SHEAVES. We shall now consider the general properties of Π-invertible sheaves in some detail, following Skornyakov.

8.11.a. Functors $u^{\pm}$. Let $\mathcal{S}$ be a locally free sheaf of rank $1|1$ (for short, $1|1$-sheaf) on a superspace X. Consider two relative flag spaces

$$X_+ = F_X(1|0; \mathcal{S}), \qquad X_- = F_X(0|1; \mathcal{S}),$$

and denote by $\pi_\pm$ their projections onto X. Put $u^\pm(\mathcal{S}) = (\pi_\pm)_*(\mathcal{O}_{X_\pm})$. Clearly, $u^\pm$ are functors on the category of $1|1$-sheaves and isomorphism. There are canonical isomorphisms $u^+(\Pi\mathcal{S}) = u^-(\mathcal{S})$ and exact sequences

$$0 \to \mathcal{O}_X \to u^\pm(\mathcal{S}) \to \mathrm{Ber}(\mathcal{S})^{\pm 1} \to 0.$$

In fact, the embedding of $\mathcal{O}_X$ corresponds to the projection $X_\pm \to X$. The quotient can be calculated in the following way.

Denote by $F(X)$ the set of isomorphism classes of $1|1$-sheaves. By general formalism, it can be identified with the cohomology set $H^1(X, \mathrm{GL}(1|1; \mathcal{O}_X))$. Functors $u^\pm$ define endomorphisms of this set. They are induced by the following idempotent endomorphisms of $\mathrm{GL}(1|1)$:

$$u^+ : \begin{pmatrix} a & b \\ c & d \end{pmatrix} = A \to \begin{pmatrix} \mathrm{Ber}(A) & b/d \\ 0 & 1 \end{pmatrix};$$
$$u^- : \begin{pmatrix} a & b \\ c & d \end{pmatrix} = A \to \begin{pmatrix} 1 & 0 \\ c/a & \mathrm{Ber}(A)^{-1} \end{pmatrix}.$$

If we restrict these endomorphisms $\mathrm{SL}(1|1)$, we obtain two morphisms of this group onto $\mathbb{G}_a^{0|1}$ defined by the point-functor $M \to \Gamma(\mathcal{O}_{M,1})$ (look at the corner elements). Putting $u = u^+ - u^-$, we get an exact sequence

$$1 \to \mathbb{G}_m^{1|1} \xrightarrow{\alpha} \mathrm{SL}(1|1) \xrightarrow{u} \mathbb{G}_a^{0|1} \to 1.$$

where

$$\alpha(b_0 + b_1) = \begin{pmatrix} b_0 & b_1 \\ b_1 & b_0 \end{pmatrix}.$$

8.11.b. An Odd Chern Class. The set of isomorphism classes $\mathrm{Pic}_\Pi(X)$ of $1|1$-sheaves with an odd involution can be identified with $H^1(X, \mathcal{O}_X^*) = H^1(X, \mathbb{G}_m^{1|1}(\mathcal{O}_X))$. The composed morphism $u^+\alpha : \mathbb{G}_m^{1|1} \to \mathbb{G}_a^{0|1}$ defines a characteristic class

$$c : \mathrm{Pic}_\Pi(X) \to H^1(X, \mathcal{O}_{X,1}).$$

The canonical embedding $\mathbb{G}_m \to \mathbb{G}_m^{1|1}$ induces a map $\mathrm{Pic}(X) \to \mathrm{Pic}_\Pi(X)$ corresponding to $\mathcal{L} \to (\mathcal{L} \oplus \Pi\mathcal{L}$, *the interchange of summands*). On such sheaves c is trivial.

8.11.c. Composition. Let $(\mathcal{S}, p)$ and $(\mathcal{S}', p')$ be two Π-invertible sheaves. Put $q = i(p \otimes p')$. It is an even involution of $\mathcal{S} \otimes \mathcal{S}'$. Let $(\mathcal{S} \otimes \mathcal{S}')_\pm$ be defined by proper values ± 1 of q. They are locally free $1|1$-sheaves with trivial Berezinians. Moreover,

$$u^{\varepsilon}((\mathcal{S}\otimes\mathcal{S}')_{\eta}) = c(\mathcal{S}') + i\varepsilon\eta c(\mathcal{S}); \quad \varepsilon, \eta \in \{+,-\}.$$

This can be verified by a straightforward calculation.

8.11.d. Existence of an Odd Symmetry. Since the structure group of a Π-invertible sheaf is reduced to a subgroup of SL, its Berezinian is trivial, i.e., if is isomorphic to $\Pi\mathcal{O}_X$. A change of trivialization of $\mathrm{Ber}(\mathcal{S})$ corresponds to the action of $H^0(X, \mathcal{O}^*_{X,0})$ on the fibers of the map $H^1(X, \mathrm{SL}(1|1, \mathcal{O}_X)) \to H^1(X, \mathrm{GL}(1|1, \mathcal{O}_X))$. If $[S]$ is the class of $\mathcal{S}$ in the first group, a new obstruction to the existence of an odd involution on $\mathcal{S}$ compatible with a given trivialization of the Berezinian is $u(\mathcal{S})$. Under the action of the group of invertible functions, this obstruction behaves as follows: $u^{\pm}(fs) = f^{\pm 1}u^{\pm}(s)$. Thus, we get the following result.

Let $\mathcal{S}$ be a $1|1$-sheaf such that $\mathrm{Ber}(\mathcal{S}) \simeq \Pi\mathcal{O}_X$. Then it admits an odd involution iff for some invertible function f on X, we have $f^2u^{+}(\mathcal{S}) = u^{-}(\mathcal{S})$.

Applying this in the situation of (iii), we see that an odd involution on $(\mathcal{S}\otimes\mathcal{S}')_{\pm}$ exists iff for some invertible f we have

$$(f^2+1)c(\mathcal{S}) = (f^2-1)c(\mathcal{S}').$$

8.11.e. Splitting of Π-Invertible Sheaves. By definition, a Π-invertible sheaf splits, if it is isomorphic to ($\mathcal{L}\oplus\Pi\mathcal{L}$, the *interchange of summands*). This is equivalent to the further reducibility of the structure group from $\mathbb{G}_m^{1|1}$ to $\mathbb{G}_m$. The subgroup $\mathbb{G}_m$ is central. It follows that the exact sequence of pointed sets H^1 corresponding to $1 \to \mathbb{G}_m \to \mathbb{G}_m^{1|1} \to \mathbb{G}_a^{0|1} \to 1$ can be extended so that we obtain an exact sequence

$$\mathrm{Pic}(X) \to \mathrm{Pic}_{\Pi}(X) \xrightarrow{c} H^1(X, \mathcal{O}_{X,1}) \xrightarrow{\partial} H^2(X, \mathcal{O}^*_{X,0}).$$

The set of obstructions to splitting is $\mathrm{Im}(c) = \mathrm{Ker}(\partial)$.

Let us show that $\partial(\gamma) = \exp(\gamma\cup\gamma)$, where $\cup : H^1(\mathcal{O}_1)\times H^1(\mathcal{O}_1) \to H^2(\mathcal{O}_0)$ and exp is induced by $\mathcal{O}_{X,0} \to \mathcal{O}^*_{X,0} : f \to e^f$. In fact, if γ is given by a Čech cocycle $(\{\gamma_{ij}\})$, one easily sees that

$$\partial(\{\gamma_{ij}\}) = \{1-\gamma_{jk}\gamma_{ik}+\gamma_{jk}\gamma_{ij}-\gamma_{ik}\gamma_{ij}\} = \{\exp(-\gamma_{jk}\gamma_{ik}+\gamma_{jk}\gamma_{ij}-\gamma_{ik}\gamma_{ij})\}.$$

Notice also that in the category of algebraic varieties,

$$\mathrm{Ker}(\exp) = \mathrm{Im}(H^2(X_{\mathrm{red}}, \mathbb{Z}) \to H^2(X, \mathcal{O}_{X,0}))$$

consists of obstructions to the realization of 2-cohomology classes by the Cartier divisors. Thus, obstructions to splitting Π-invertible sheaves are their square roots.

8.11.f. Splitting of Arbitrary $1|1$-Sheaves. As in (v), we want to calculate $\mathrm{Ker}(\partial) = \mathrm{Im}(\lambda)$ in the exact sequence

$$\mathrm{Pic}(X) \to H^1(X, \mathrm{GL}(1|1, \mathcal{O}_X)) \xrightarrow{\lambda} H^1(X, \mathbb{H}(\mathcal{O}_X)) \xrightarrow{\partial} H^2(X, \mathcal{O}_X^*),$$

where $\mathbb{H}$ is a semidirect product of $\mathbb{G}_m$ by $\mathbb{G}_a^{0|1} \times \mathbb{G}_a^{0|1}$ corresponding to the action

$$f(\beta, \beta') = (f\beta, f^{-1}\beta'),$$

and λ is defined on the coefficients by

$$\lambda\begin{pmatrix} a & b \\ c & d \end{pmatrix} = \left\{-\frac{b}{a}, -\frac{c}{d}; \mathrm{Ber}\begin{pmatrix} a & b \\ c & d \end{pmatrix}\right\}.$$

In order to describe $\mathrm{Ker}(\partial)$, we need one more exact sequence of non-commutative cohomology derived from the definition of $\mathbb{H}$:

$$H^0(X, \mathcal{O}_{X,0}^*) \to H^1(X, \mathcal{O}_{X,1} \oplus \mathcal{O}_{X,1}) \xrightarrow{\mu} H^1(X, \mathbb{G}_m(\mathcal{O}_X)) \to \mathrm{Pic}(X).$$

Denote by ψ the composed map

$$\partial \circ \mu : H^1(X, \mathcal{O}_{X,1} \oplus \mathcal{O}_{X,1}) \to H^2(X, \mathcal{O}_X^*).$$

As in (v), we can show that

$$\psi(\beta \oplus \gamma) = \exp(\beta \cup \gamma).$$

On the other hand,

$$\mathrm{Ker}(\partial) \supseteq \mu(\mathrm{Ker}(\psi)) \simeq \mathrm{Ker}(\psi)/H^0(X, \mathcal{O}_{X,0}^*).$$

Let us apply the previous results in the case when

$$H^0(X, \mathcal{O}_X) = \mathbb{C}; \qquad \mathrm{Pic}(X) = H^1(X, \mathcal{O}_{X,0}^*) = \{1\}.$$

Notice that the last condition in pure even geometry would be highly unusual for proper reduced algebraic varieties; it would imply, in particular, that X is nonprojective. However, both conditions are valid for superprojective spaces $\mathbb{P}_{\Pi}^{n|n}$ and, more generally, for Π-flag spaces $Q(n)/P$ (cf. Chapter 3).

Put

$$W^1 = H^1(X, \mathcal{O}_{X,1}); \qquad W^2 = H^2(X, \mathcal{O}_{X,0}); \qquad L = H^2(X, \mathbb{Z}).$$

As we have remarked, L consists of obstructions to the realization of 2-cocycles by Cartier divisors. By assumption, it is embedded W^2. From Section 8.11.e, we see that

$$\mathrm{Pic}_{\Pi}(X) = \{a \in W^1 \mid a^2 \in L\}.$$

On the other hand, from (vi):

$$H^1(X, \mathrm{GL}(1|1, \mathcal{O}_X)) = \{(\beta, \gamma) \in W^1 \oplus W^1 \mid \beta\gamma \in L\}/\mathbb{C}^*.$$

The image of $\mathrm{Pic}_{\Pi}(X)$ in this set is

$$\{(\beta, \gamma) \in W^1 \oplus W^1 \mid \exists f \in \mathbb{C}^* \quad \text{s.t.} \quad f^2\beta = \gamma \quad \text{and} \beta\gamma \in L\}/\mathbb{C}^*.$$

8.12. PROBLEM. Is it possible to develop a parallel theory of algebraic curves in supergeometry in which a basic example of the curve of genus zero would be $\mathbb{P}_{\Pi}^{1|1}$ with its automorphism group instead of $\mathbb{P}^{1|1}$ with its structure group reduced to $\mathrm{SpO}(2|1)$?

CHAPTER 3

Flag Superspaces and Schubert Supercells

1. Classical Supergroups and Flag Superspaces

1.1. CLASSICAL SUPERGROUPS. In this section, we shall first introduce classical supergroups and their flag spaces and then shall define combinatorial invariants that will be used for enumeration of Schubert supercells. The now standard approach based on the properties of Weyl groups has some drawbacks in our situation, and we therefore fall back upon the more classical notion of "relative position" of a pair of flags.

We start with recalling and completing some constructions discussed in [Ma1]. A supergroup of the type SL, OSp, $\Pi\,\mathrm{Sp}$, or Q is defined by its standard representation space $T \simeq \mathbb{C}^{m|n}$ and an invariant. For SL, the invariant is an arbitrary nonzero element of $\mathrm{Ber}(T)$; for OSp, it is a nondegenerate even symmetric form $b : T \to T^*$; for $\Pi\,\mathrm{Sp}$, it is a nondegenerate odd alternating form $b : T \to T^*$; finally, for Q, it is an odd involution $p : T \to T$, $p^2 = \mathrm{id}$. We fix b or p; its existence in cases $\Pi\,\mathrm{Sp}$ and Q implies $m = n$. The groups $\mathrm{OSp}(2r \mid 2s)$ have two connected components; the rest of them are connected.

1.2. FLAG SPACES. Consider the set ${}^{\mathrm{SL}}\mathbf{I}$ of all sequences $I = (\delta_1, \dots, \delta_r)$, $\delta_i = p_i \mid q_i$, $\sum_{i=1}^r \delta_i = m \mid n$, $r \leq m + n$. Choose an I and a superscheme S. Put $T_S = T \otimes \mathcal{O}_S$. A *flag of type I in T_S* is a sequence $0 = \mathcal{S}_0 \subset \mathcal{S}_1 \subset \cdots \subset \mathcal{S}_{r-1} \subset \mathcal{S}_r = T_S$ of locally free locally direct subsheaves such that $\mathrm{rk}(\mathcal{S}_i) - \mathrm{rk}(\mathcal{S}_{i-1}) = \delta_i$ for all $1 \leq i \leq r$. Hence, $\mathrm{rk}(\mathcal{S}_i) = d_i(I) = \delta_1 + \cdots + \delta_i$.

Denote by ${}^{\mathrm{SL}}F_I$ the functor that associates with a superscheme S the set of all flags of the type I in T_S.

The functors ${}^{G}F_I$ for other classical groups G consist of the flags subject to the following restrictions:

$$\text{for} \quad \mathrm{OSp} \text{ and } \Pi\,\mathrm{Sp} : (b \otimes \mathrm{id}_S)(\mathcal{S}_1) = \mathcal{S}_{r-1}^{\perp};$$
$$\text{for} \quad Q : (p \otimes \mathrm{id}_S)(\mathcal{S}_1) = \mathcal{S}_i; \quad i = 0, \dots, r.$$

Here the orthogonal complement is taken with respect to b.

From the definition, it follows that ${}^GF_I$ is nonempty only if I satisfies the following conditions:

$$\delta_i(I) = \begin{cases} \delta_{r+1-i}(I) & \text{for} \quad \mathrm{OSp}; \\ \delta^c_{r+1-i}(I) & \text{for} \quad \Pi\,\mathrm{Sp}; \\ \delta^c_i(I) & \text{for} \quad Q, \end{cases}$$

where $(p \mid q)^c = q \mid p$. Let ${}^G\mathbf{I}$ be the set of such G-types.

As is shown in [Ma1], all of the functors ${}^GF_I$ are representable by supermanifolds.

1.3. Complete Flags. A G-flag is *complete* if it has a maximal possible length r. Denote by ${}^G\mathbf{I}_c$ the set of types of complete G-flags. Clearly, it admits the following description.

$$\begin{aligned} {}^{\mathrm{SL}}\mathbf{I}_c &= \{(\delta_1, \dots, \delta_{m+n}) \mid \delta_i = 1 \mid 0 \text{ or } 0 \mid 1, \sum_i \delta_i = m \mid n\}; \\ {}^{\mathrm{OSp}}\mathbf{I}_c &= \{(\delta_1, \dots, \delta_{m+n}) \in {}^{\mathrm{SL}}\mathbf{I}_c \mid \delta_i = \delta_{m+n+1-i}\}; \\ {}^{\Pi\,\mathrm{Sp}}\mathbf{I}_c &= \{(\delta_1, \dots, \delta_{m+n}) \in {}^{\mathrm{SL}}\mathbf{I}_c \mid \delta_i = \delta_{m+n+1-i}\}; \\ {}^{Q}\mathbf{I}_c &= \{(1 \mid 1, \dots, 1 \mid 1) \text{ of length } m\}. \end{aligned}$$

Flag spaces of complete types ${}^GF_I$ are connected except for the case $\mathrm{OSp}(2r \mid 2s)$, where they consist of two connected components. We shall put

$${}^GF = \bigsqcup_{I \in {}^GI_c} {}^GF_I$$

and shall consider this space as an analog of classical spaces $B\backslash G$, B a Borel subgroup.

1.4. Relative Position of a Pair of Complete Flags. Let $\mathcal{S}'$, $\mathcal{S}''$ be two complete G-flags of types I and J, respectively, in T_S. We shall say that $\{\mathcal{S}', \mathcal{S}''\}$ is *well positioned* if for all i, j, locally on S, there exists a direct decomposition $T_S = \mathcal{T}_0 \oplus \mathcal{T}_1 \oplus \mathcal{T}_2 \oplus \mathcal{T}_3$ such that $\mathcal{S}'_i = \mathcal{T}_0 \oplus \mathcal{T}_1$, $\mathcal{S}''_j = \mathcal{T}_0 \oplus \mathcal{T}_2$, with the ranks of $\mathcal{T}_K$ depending only on i, j. In particular, $\mathcal{S}'_i \cap \mathcal{S}''_j$ and $\mathcal{S}'_i + \mathcal{S}''_j$ are locally direct subsheaves of T_S of constant rank. Equivalently, for all $i, j, T_S/(\mathcal{S}'_i + \mathcal{S}''_j), \mathcal{S}'_i/(\mathcal{S}'_i \cap \mathcal{S}''_j)$ and $\mathcal{S}''_j/(\mathcal{S}' \cap \mathcal{S}''_j)$ are locally free of finite rank.

The *type of relative position* of a well-positioned pair $\mathcal{S}', \mathcal{S}''$ is the matrix with entries

$$d_{ij} = \mathrm{rk}(\mathcal{S}'_i \cap \mathcal{S}''_j), \qquad 0 \le i, j \le r,$$

where $r = m + n$ for $G = \mathrm{SL}, \mathrm{OSp}, \Pi\,\mathrm{sp}$; $r = m$ for $G = Q$.

1.5. LEMMA. *(a) The matrix (d_{ij}) has the following properties:*

$$d_{0j} = 0 \mid 0; \quad d_{i0} = 0 \mid 0; \quad d_{rj} = d_j(J); \quad d_{ir} = d_i(I);$$
$$0 \mid 0 \le d_{ij} - d_{i-1,j} \le \delta_i(I); \quad 0 \mid 0 \le d_{ij} - d_{i,j-1} \le \delta_j(J);$$
$$d_{ij_0} \ne d_{i-1,j_0} \Rightarrow d_{ij} \ne d_{i-1,j} \quad \textit{for} \quad j \ge j_0;$$
$$d_{i_0j} \ne d_{i_0,j-1} \Rightarrow d_{ij} \ne d_{i,j-} \quad \textit{for } i \ge i_0;$$

where $a \mid b \le a' \mid b'$ means that $a \le a'$ and $b \le b'$.

(b) If $G = \mathrm{OSp}, \Pi\,\mathrm{sp}$, the matrix (d_{ij}) has the following additional symmetry properties:

$$\mathrm{OSp}: d_{ij} = d_{m+n-i,m+n-j} - m \mid n + d_{m+n-i}(I) + d_{m+n-j}(J);$$
$$\Pi\,\mathrm{Sp}: d_{ij} = d^c_{2m-i,2m-j} - m \mid m + d^c_{2m-i}(I) + d^c_{2m-j}(J).$$

1.6. THE FLAG WEYL GROUPS. We define the action of S_{m+n} on ${}^{\mathrm{SL}}\mathbf{I}_c$ by $\delta_i(wI) = \delta_{w^{-1}(i)}(I)$ and then introduce the flag Weyl groups ${}^GW$ by the following prescription:

(a) S_{m+n} for $G = \mathrm{SL}(m \mid n)$;

(b) $\{w \in^{\mathrm{SL}(m|n)} W \mid w({}^G\mathbf{I}_c) = {}^G\mathbf{I}_c\}$ for $G = \mathrm{OSp}(m \mid n)$ or $\Pi\,\mathrm{Sp}(m)$ (then $m = n$);

(c) S_m for $G = Q(m)$, with an obvious action on ${}^Q\mathbf{I}_c$.

We shall see that this notion is well suited for the investigation of the geometry of Weyl supercells, but it has some peculiar properties in comparison with the classical Weyl groups. For example, as Penkov and Skornyakov remarked, for $\mathrm{OSp}(2 \mid 2)$ we have $W = S_2 \ltimes (\mathbb{Z}_2)^2$ and for $\mathrm{SL}(1 \mid 2)$ we have $W = S_3$, although the corresponding Lie superalgebras are isomorphic.[1] The intuition behind this notion is that we want to include into our group the reflections with respect to odd roots, or, in terms of the fundamental representation, to interchange even and odd vectors, and this cannot be realized by a morphism of superspaces.

1.7. LEMMA. *The flag Weyl group ${}^GW$ for $\mathrm{OSp}(m \mid n)$ and $\Pi\,\mathrm{Sp}(m)$ is isomorphic to the semidirect product $S_t \ltimes (\mathbb{Z}_2)^t, t = [(m+n)/2]$, where $m = n$ for $\Pi\,\mathrm{Sp}$, and the semidirect product is constructed with respect to the natural action of S_t by permutations of factors of $(\mathbb{Z}_2)^t$.*

Proof. The types of complete flags are symmetric with respect to the middle point (for $G = \Pi\,\mathrm{Sp}$, the symmetry involves the operation $(|)^c$).

---

[1]As a referee pointed out, problems occur already in the pure even case, with $A_3 = D_3$.

Clearly, a permutation is compatible with such a symmetry iff it can be represented as a composition of the following elementary permutations:

(a) any permutation of the left half followed by the symmetric permutation of the right half;

(b) any transposition of a mirror symmetric pair of elements.

We shall now show how to characterize the types of relative position of complete flags with the help of the flag Weyl groups.

1.8. PROPOSITION. *There is a natural bijection between the following sets:*

(a) types of relative position of complete G-flags;

(b) matrices (d_{ij}) *with the properties stated in Lemma 1.5.*

(c) triples $\{(I, J, w) \mid I, J \in {}^G\mathbf{I}_c, w \in {}^GW, J = w(I)\}$.

Proof. We start with the case SL. Suppose first that a triple (I, J, w) of the type (c) is given. Then we can construct a corresponding pair of flags in the following way. Choose a basis $\{e_{k,I}\}$ in $T^{m|n}$ such that $\dim e_{k,I} = \delta_k(I)$. Put

$$S_{I,i} = \langle e_{1,I}, \ldots, e_{i,I}\rangle;$$
$$S_{J,j} = \langle e_{w^{-1}(1),I}, \ldots, e_{w^{-1}(j),I}\rangle.$$

Denote by $(d_{ij,w,I,J})$ the type of relative position of these flags. This gives a map (c) $\to$ (a) $\to$ (b).

In order to reconstruct (I, J, w) from (d_{ij}), let us first look at the case $d_{ij} = d_{ij,w,I,J}$. Consider pairs of unequal vertical neighbors in this matrix and mark the leftmost members of these pairs in each pair of consecutive rows.

The first column contains precisely one pair of unequal neighbors ("jump"), say,

$$d_{k_1,1} \neq d_{k_1-1,1} \Leftrightarrow e_{k_1,I} = e_{w^{-1}(1),I} \Leftrightarrow w^{-1}(1) = k_1.$$

The second column contains a jump between the entries of the k_1th and $(k_1 - 1)$-th lines and exactly one more jump:

$$d_{k_2,2} \neq d_{k_2-1,2}, \quad k_2 \neq k_1 \Leftrightarrow e_{k_2,I} = e_{w^{-1}(2),I} \Leftrightarrow w^{-1}(2) = k_2.$$

This pattern continues. In this way, we see that I = the last column of (d_{ij}), J = the last line of d_{ij}, and w is defined by $w(k_i) = i$.

Now notice that this procedure can be repeated for any matrix (d_{ij}) satisfying the conditions of Lemma 1.5 (a). In fact, use induction on the column number. In the first column, the zeroth element is $0 \mid 0$; and later on, we have several jumps by $0 \mid 1$ or $1 \mid 0$; but the last element is $d_1(J) = \delta_1(J) = 0 \mid 1$

or $1 \mid 0$, so that there is exactly one jump. To the right of it, the jumps will be repeated. Assume that we have already proved that in the $(k-1)$-th column there are $(k-1)$ jumps that repeat themselves to the right. Since the kth column starts with $0 \mid 0$ and ends with $d_k(J)$, and the jumps are all of the type $0 \mid 1$ or $1 \mid 0$, there should be all in all k jumps, of which $(k-1)$ descend from the previous ones. Hence, the only new jump is leftmost in its line.

We must convince ourselves now that if (I, J, w) is reconstructed from (d_{ij}), then $w(J) = I$, that is $\delta_i(J) = \delta_{w^{-1}(i)}(I) = \delta_{k_i I}$. Notice first that if $d_{ij} \neq d_{i-1,j}$ and $d_{i,j-1} = d_{i-1,j-1}$, then $d_{ij} \neq d_{i,j-1}$ and $d_{i-1,j-1} = d_{i-1,j}$. Now, by construction

$$d_{k_i,i} \neq d_{k_i-1,i} \quad \text{and} \quad d_{k_i,i-1} = d_{k_i-1,i-1};$$

therefore,

$$d_{k_i,i} \neq d_{k_i,i-1} \quad \text{and} \quad d_{k_i-1,i} = d_{k_i-1,i-1};$$

and finally,

$$\delta_{k_i}(I) = d_{k_i,i} - d_{k_i-1,i} = d_{k_i,i} - d_{k_i,i-1} = \delta_i(J).$$

Clearly, the map (b) $\to$ (c) constructed in this way is inverse to the map (c) $\to$ (b). By using it, and by using as an intermediary the matrix of the relative position of two flags, we can also construct (a) $\to$ (c). This finishes our proof for SL.

For $G = Q$, the proof follows the same pattern as for $\mathrm{SL}(m \mid 0)$.

For $G = \mathrm{OSp}(m \mid n), \Pi\,\mathrm{Sp}(m)$, one should only take into account the Weyl symmetry. If we start with a (d_{ij}) with G-symmetry, we first get $(I, J, w); I, J \in {}^{\mathrm{SL}}\mathbf{I}_c, w \in {}^{\mathrm{SL}}W, J = w(I)$. From $I = (d_{i,m+n})$ and $J = (d_{m+n,j})$, it follows that $I, J \in {}^{G}I_c$. From the properties 1.5 (b), one derives that $w \in {}^{G}W$. In this way, we obtain a map (b) $\to$ (c). The rest of the proof is the same as for SL.

2. Schubert Supercells

2.1. Reduced Cells. We fix a classical supergroup G and put $F = {}^{G}F$, $W = {}^{G}W$. For an element $w \in W$, denote by $d_{ij,w}$ a function on $\bigsqcup_{J=w(I)} F_I \times F_J$ with constant values on $F_I \times F_J$:

$$d_{ij,w|F_I \times F_J} = d_{ij,w,IJ}$$

(this superdimension was defined in the proof of Proposition 1.8). Let $\mathcal{S}$ be the tautological flag on F and $p_{1,2} : F \times F \to F$ projections. Put $|Y_w|(\mathbb{C}) =$ the set of all $\mathbb{C}$-points of $\bigsqcup_{J=w(I)} F_I \times F_J$ over which for all $i, j, \mathcal{S}_i', \mathcal{S}_j''$ are well positioned and $\dim(p_1^*\mathcal{S}_i \cap p_2^*\mathcal{S}_j) = d_{ij,w}$.

Clearly, it is the set of $\mathbb{C}$-points of a locally closed subset $|Y_w| \subset (F \times F)_{\text{red}}$. It follows from Proposition 1.8 that the union of these subsets coincides with $(F \times F)_{\text{red}}$.

We shall introduce on Y_w the structure of a locally closed subsupermanifold by using the technique of flattening stratifications (cf. [Mu]).

2.2. THEOREM. *(a) On each $|Y_w|$, there exists a canonical structure of a locally closed subsuperscheme $Y_w \subset F \times F$ such that the disjoint union of Y_w is the flattening stratification for the family of sheaves* $\mathcal{F}(\mathcal{S}', \mathcal{S}'') = \{\mathcal{T}_S/(p_1^*\mathcal{S}_i' = p_2^*\mathcal{S}_j''), \mathcal{S}_i'/(\mathcal{S}_i' \cap \mathcal{S}_j''), \mathcal{S}_j''/(\mathcal{S}_i' \cap \mathcal{S}_j'')\}$.

This means that a morphism $g : S \to F \times F$ (S noetherian) can be represented as a composition of $g' : S \to \bigsqcup Y_w$ and the canonical inclusion iff it has the following property: all sheaves from $\mathcal{F}(\mathcal{S}', \mathcal{S}'')$ are flat $\mathcal{O}_S$-modules.

(b) $p_2 : Y_w \to F$ is a relative affine superspace.

We shall first consider the case of a relative projective superspace.

2.3. PROJECTIVE SPACE DECOMPOSITION. Let X be a fixed noetherian superscheme, $\mathcal{T}$ a locally free sheaf of rank $m|n$ on it, and $\mathcal{S}$ a fixed complete flag in $\mathcal{T}$. It defines in the relative projective space $\mathbb{P}_X(1 \mid 0; \mathcal{T})$ an ascending chain

$$\emptyset \subseteq \mathbb{P}_X(1 \mid 0; \mathcal{S}_1) \subseteq \mathbb{P}_X(1 \mid 0; \mathcal{S}_2) \subseteq \cdots \subseteq \mathbb{P}_X(1 \mid 0; \mathcal{T}).$$

Let Z_k be the open subsuperscheme in $\mathbb{P}_X(1 \mid 0; \mathcal{S}_k)$ with support complementary to that of the previous projective space in the chain. (It is empty if $\dim \mathcal{S}_k/\mathcal{S}_{k-1} = 0 \mid 1$.)

2.4. PROPOSITION. *(a) $\bigsqcup Z_k$ is the relative flattening stratification for the family of sheaves $\mathcal{F}(\mathcal{O}(-1), \mathcal{S})$. (Relative refers to the category of X-superschemes and X-morphisms.)*

(b) Z_k is a relative affine space over X of dimension $\operatorname{rk}(\mathcal{S}_k) - 1 \mid 0$.

We leave the proof to the reader.

2.5. COROLLARY. *In the situation of Section 2.3–2.4, there exists a relative flattening stratification $\bigsqcup Y_k$ of $\mathbb{P}_X(1 \mid 0; \mathcal{T}) \times_X F_X$ for the family of sheaves $\mathcal{F}(p_1^*\mathcal{O}(-1), \mathcal{S})$. The fiber of $p_2 : \bigsqcup Y_k \to F_X$ over an X-point of F_X represented by the flag $\mathcal{S}$ in $\mathcal{T}$ equals $\bigsqcup Z_k$.*

2.6. *Proof of* 2.2. We can repeat the proof of the simplest particular case of the flattening decomposition theorem (cf. [Mu], Section 6.5, Proposition). For a sheaf $\mathcal{S}$ in $\mathcal{F}(\mathcal{S}', \mathcal{S}'')$ and a closed point $x \in |Y_w|$, we chose a minimal local exact sequence (in a neighborhood of x) $\mathcal{P}'' \xrightarrow{f} \mathcal{P}' \to \mathcal{S} \to 0$, where $\mathcal{P}', \mathcal{P}''$ are locally free, and then define Y_w in this neighborhood by $f = 0$. The second part follows from the corresponding statement in the pure even case.

Finally, notice that for a flag $\mathcal{S}$ and a connected component X of the flag space, in general there is no relative position w for which the set $\{\mathcal{S}' \mid \mathcal{S}' \in X$, well positioned with respect to $\mathcal{S}$ and with the position $w\}$ is open and dense in X.

2.7. PROPOSITION. *Consider two complete G-flags in T_S (notations in 1.1 and 1.2) $\mathcal{S}', \mathcal{S}''$, with the following properties:*

(a) $\{\mathcal{S}', \mathcal{S}''\}$ is well positioned.

(b) The type of relative position is (d_{ij}).

Let $\bar{\mathcal{S}}', \bar{\mathcal{S}}''$ be another well-positioned pair of G-flags of the same type and the same type of relative position. Then every point of S has an open affine neighborhood $U = \operatorname{Spec}(A)$ such that there exists an element $g \in G(A)$ transforming the first pair into the second one.

Proof. We start again with the case $G = \mathrm{SL}$. Localizing on S, we may and will assume that all sheaves

$$\mathcal{S}'_i, \mathcal{S}''_j, \bar{\mathcal{S}}'_i, \bar{\mathcal{S}}''_j, \quad \mathcal{S}'_i \cap \mathcal{S}''_j, \bar{\mathcal{S}}'_i \cap \bar{\mathcal{S}}''_j$$

are free direct subsheaves of T_S.

Then we can construct $g \in G(A)$ explicitly by giving its action on a basis of sections of T_S compatible with the given pair of flags.

Start by choosing an isomorphism $\mathcal{S}_1 \to \bar{\mathcal{S}}_1$. Then this map transforms the degenerate flag $\mathcal{S}'_1 \cap \mathcal{S}''$ into the degenerate flag $\bar{\mathcal{S}}' \cap \bar{\mathcal{S}}''$ because (d_{1j}) for both pairs of flags coincide.

Assume now that g is already defined on $\mathcal{S}'_k$, transforms the first k components of $\mathcal{S}'$ into the first k components of $\bar{\mathcal{S}}'$, and $\mathcal{S}'_k \cap \mathcal{S}''$ into $\bar{\mathcal{S}}'_k \cap \bar{\mathcal{S}}''$. Let us extend g onto $\mathcal{S}'_{k+1}$ in such a way that the same properties hold for $k+1$.

Denote by j_0 the smallest j with $d_{k+1,j} \neq d_{kj}$. Choose in $\mathcal{S}'_{k+1} \cap \mathcal{S}''_{j_0}$ an element e_{k+1} complementing $\mathcal{S}'_k \cap \mathcal{S}''_{j_0}$ and choose an element $\bar{e}_{k+1}$ with the same properties for the second pair of flags. It is possible in view of a dimension count. Extend g to e_{k+1} by transforming it to $\bar{e}_{k+1}$. Clearly, g transforms the first $k+1$ components of $\mathcal{S}'$ into those of $\bar{\mathcal{S}}'$.

Let us check now that $\mathcal{S}'_{k+1} \cap \mathcal{S}''$ transforms itself into $\bar{\mathcal{S}}'_{k+1} \cap \bar{\mathcal{S}}''$. By dimension count, for $j < j_0$, we have $\mathcal{S}'_{k+1} \cap \mathcal{S}''_j = \mathcal{S}'_k \cap \mathcal{S}''_j$ and similarly for the second pair, so that

$$g(\mathcal{S}'_{k+1} \cap \mathcal{S}''_j) = \bar{\mathcal{S}}'_{k+1} \cap \bar{\mathcal{S}}''_j$$

by inductive assumption; the same is true for $j = j_0$ by construction. Finally, for $j > j_0$,

$$\mathcal{S}'_{k+1} \cap \mathcal{S}''_j = \langle e_{k+1} \rangle \oplus (\mathcal{S}'_k \cap \mathcal{S}''_j),$$

and we know that

$$e_{k+1} \in \mathcal{S}'_{k+1} \cap \mathcal{S}''_{j_0} \subset \mathcal{S}'_{k+1} \cap \mathcal{S}''_j;$$
$$e_{k+1} \notin \mathcal{S}''_k \supset \mathcal{S}'_k \cap \mathcal{S}''_j.$$

Hence,

$$g(\mathcal{S}'_{k+1} \cap \mathcal{S}''_j) = \langle \bar{e}_{k+1} \rangle \oplus \bar{\mathcal{S}}'_k \cap \bar{\mathcal{S}}''_j = \bar{\mathcal{S}}'_{k+1} \cap \bar{\mathcal{S}}''_j.$$

At the end of this construction, we necessarily get an element g' from $\mathrm{GL}(A)$. In order to achieve $\mathrm{Ber}(g) = 1$, it suffices to correct g' at the last step, say, by $g(e_k) = (\mathrm{Ber}(g))^{\pm 1}\bar{e}_k$ ($\pm$ at the exponent depends on the parity of e_k).

If $G = \mathrm{OSp}$ or $\Pi\,\mathrm{Sp}$, we carry out the same inductive construction up to $[(m+n+1)/2]$, where $m \mid n = \dim T$. Afterwards, we choose $e_k, \bar{e}_k$ in such a way that $b(e_k, e_{m+n+1-k}) = 1, b(e_k, e_j) = 0$ for the rest of j, and similarly for $\bar{e}_k$. Then $g \in G(A)$, since g conserves the Gram matrix of b.

Finally, for $G = Q$, we carry out the same inductive construction, but at each step, the dimension of $\mathcal{S}'_k$ to which we extend g is augmented by $1 \mid 1$, because in the complement to $\mathcal{S}'_k \cap \mathcal{S}''_{j_0}$ in $\mathcal{S}'_{k+1} \cap \mathcal{S}''_{j_0}$, we choose a basis $(e_{k+1}, p(e_{k+1}))$.

3. Superlength in Flag Weyl Groups

3.1. Definition. The following elements of the flag Weyl groups will be called *basic reflections*:

(a) $G = \mathrm{SL}: \ \sigma_i = (i, i+1)$ (a transposition of neighbors in $\{1, 2, \dots, m+n\}$).

(b) $G = \mathrm{OSp}$ or $\Pi\,\mathrm{Sp}$:

$$\sigma_i = (i, i+1)(m+n+1-i, m+n-i), \quad i+1 \leq [(m=n)/2];$$
$$\tau_i = (l, m+n+1-l), \quad l = [(m+n)/2].$$

(c) $G = Q: \ \sigma_i = (i, i+1) \in S_m$.

3.2. Definition. The superlength of an element $w \in {}^G W$ is defined by the following prescription.

(a) Let $I, J \in {}^G I_c, \sigma \in {}^G_W$ be a basic reflection and $J = \sigma(I)$. Then, for $\sigma = \sigma_i$,

$$l_{IJ}(\sigma) = \begin{cases} 1 \mid 0, & \text{if } I = J;\ G = \mathrm{SL}, \mathrm{OSp}, \Pi\,\mathrm{Sp}. \\ 0 \mid 1, & \text{if } I \neq J;\ G = \mathrm{SL}, \mathrm{OSp}, \Pi\,\mathrm{Sp}. \\ 1 \mid 1, & \text{if } G = Q. \end{cases}$$

For $\sigma = \tau_l$ and $\delta_l(J) = 1 \mid 0$;

$$l_{IJ}(\sigma) = \begin{cases} 1 \mid 0, & \text{if } G = \mathrm{OSp}(2r+1 \mid 2s). \\ 0 \mid 0, & \text{if } G = \mathrm{OSp}(2r \mid 2s). \\ 0 \mid 1, & \text{if } G = \Pi\,\mathrm{Sp}(m). \end{cases}$$

For $\sigma = \tau_l$ and $\sigma_l(J) = 0 \mid 1$;

$$l_{IJ}(\sigma) = \begin{cases} 1 \mid 1, & \text{if } G = \mathrm{OSp}(2r+1 \mid 2s). \\ 1 \mid 0, & \text{if } G = \mathrm{OSp}(2r \mid 2s). \\ 0 \mid 0, & \text{if } G = \Pi\,\mathrm{Sp}(m). \end{cases}$$

(b) For an arbitrary $w \in {}^G W$, let $w = \sigma^k \dots \sigma^1$ be a shortest decomposition of w into a product of basic reflections, $J = w(I)$. Put $I_i = \sigma^i \dots \sigma^1(I)$. Then the superlength of w is

$$l_{IJ}(w) = \sum_{i=0}^{k-1} l_{I_i, I_{i+1}}(\sigma^{i+1}),$$

while the length of w is k.

3.3. Theorem. *The notion of superlength is well defined, and*

$$\dim(Y_w \cap (F_I \times F_J)) = l_{IJ}(w) + \dim(F_I).$$

3.4. Corollary. *We have*

$$l_{IJ(w)} + \dim(F_I) = l_{IJ}(w^{-1}) + \dim(F_J).$$

This follows from the fact that the permutation of the factors F_I, F_J transforms Y_w into itself.

It is therefore interesting to understand when $\dim(F_I)$ is independent of I. Only in the case $\Pi\,\mathrm{Sp}$, there may be exceptions. In fact, by using [Ma1] and

Theorem 6.3 of Chapter 5, one can calculate the dimensions of F_I explicitly. The result is:

$$\begin{aligned}\mathrm{SL}(m \mid n): &\quad (m(m-1)/2 + n(n-1)/2 \mid mn);\\ \mathrm{OSp}(2r+1 \mid 2s): &\quad (r^2 + s^2 \mid s(2r+1));\\ \mathrm{OSp}(2r \mid 2s): &\quad (r(r-1) + s^2 \mid 2rs);\\ Q(m): &\quad (m(m+1)/2 \mid m(m+1)/1);\end{aligned}$$

finally,

$$\begin{aligned}\Pi\,\mathrm{Sp}(m): \quad (rs + r(r-1)/2 + s(s-1)/2 \mid rs\\ + r(r+1)/2 + s(s-1)/2),\end{aligned}$$

where this time, $r \mid s$ is the dimension of the maximal isotropic component of the flag of the type $I, r+s=m$.

3.5. Proof of Theorem 3.3. We shall proceed by induction on the length k of $w = \sigma^k \dots \sigma^1$.

For $k=0$, we have $w = e, J = I, l_{II}(e) = 0 \mid 0$, and it suffices to check that $Y_e \cap (F_I \times F_I) \simeq F_I$. Actually, Y_e is the diagonal of $F \times F$. In order to see this, notice that

$$d_{ij,e,II} = d_{\min(i,j)}(I).$$

On the other hand,

$$(S'_{I,i} \cap S''_{I,j}) \mid_\Delta = S'_{I,\min(i,j)} \mid_\Delta,$$

so that these sheaves are locally free locally direct subsheaves in T_Δ of ranks $d_{\min(i,j)}(I)$. Hence, if $g: S \to F \times F$ is a morphism into the diagonal, then all sheaves $g^*(S'_{I,i} \cap S''_{I,j})$ are locally free locally direct subsheaves of T_S of ranks $d_{ij,e,II}$. In the same manner, one obtains the reverse statement.

Suppose now that our theorem is already proved for all w of length k. Consider a reduced decomposition $w = \sigma^{k+1} \dots \sigma^1$ and put $w_0 = \sigma^k \dots \sigma^1, J_0 = w_0(I)$.

We shall need the following facts:

(a) If $\sigma^{k+1} = \sigma_q$ (a basic reflection defined in Section 3.1), then $w_0^{-1}(q) < w_0^{-1}(q+1)$.

(b) If $\sigma^{k+1} = \tau_l, l = [(m+n)/2]$, then $w_0^{-1}(l) < w_0^{-1}(m+n+1-l)$.

This is just a restatement in terms of permutations of the following classical observation.

Let w be an element of a classical Weyl group of one of the types A, B, C; and let γ be a positive root. If $l(w) = l(\sigma_\gamma w) - 1$, where l is the length, then $w^{-1}(\gamma)$ is positive.

Let us now introduce a new notation: if $\sigma^{k+1} = \sigma_q$, put $a = w_0^{-1}(q)$, $b = w_0^{-1}(q+1)$; if $\sigma^{k+1} = \tau_l$, put $a = w_0^{-1}(l), b = w_0^{-1}(m+n+1-l)$. As we have remarked, $a < b$. Now consider the two cases separately.

Case $\sigma^{k+1} = \sigma_q$.
From the definition of $d_{ij,w,IJ}$ in Section 1.8, it follows that

$$\text{for } i \geq b: \quad d_{iq,w,IJ} = d_{iq,w_0,IJ_0} + \delta_{q+1}(J_0) - \delta_q(J_0);$$
$$\text{for } a \leq i < b: \quad d_{iq,w,IJ} = d_{iq,w_0,IJ_0} - \delta_q(J_0)$$

and symmetric relations for OSp and Π Sp;

$$\text{for the rest of } i, j: \quad d_{ij,w,IJ} = d_{ij,w_0,IJ_0}.$$

Looking at $d_{ij,w,IJ}$ as the dimension of the intersection of two flag components, one can reinterpret these identities geometrically.

Consider natural projections $F_{J_0} \to \overline{F}$ and $F_J \to \overline{F}$ onto a superspace of incomplete flags corresponding to the deletion of the qth components $\mathcal{S}''_{J_0,q}$ and $\mathcal{S}''_{J,q}$ and their orthogonals if $G = \text{OSp}$ or $\Pi\,\text{Sp}$. Take their direct products with id_{F_I} and denote by ρ_0, ρ the resulting morphisms: $\rho_0 : F_I \times F_{J_0} \to F_I \times \overline{F}$, $\rho : F_I \times F_J \to F_I \times \overline{F}$. Then from the previous equalities it follows that

$$\rho_0(Y_{w_0}) = \rho(Y_w) = Y.$$

(Recall that Y_w is one of the components of the flattening stratification for the family of sheaves $\{\mathcal{S}'_{I,i} \cap \mathcal{S}''_{J,i}\}$). In fact, ρ_0 restricted to Y_{w_0} is an embedding; Y_w is the large cell of a relative projective superspace of dimension $l_{J_0 J}(\sigma^{k+1})$ over Y (or, for $G = Q$, of the supergrassmanian of p-symmetric $1 \mid 1$-subsheaves in a $2 \mid 2$-sheaf). This again follows from the properties of (d_{ij}) and the previous equalities rewritten in the following form, where all sheaves are locally free:

$$(3.1) \qquad \text{rk}(\mathcal{S}'_{I,i} \cap \mathcal{S}''_{J_0,q}/\mathcal{S}''_{J_0,q-1}) \mid Y_{w_0} = \begin{cases} 0 \mid 0 & \text{for } i < a; \\ \delta_q(J_0) & \text{for } i \geq a. \end{cases}$$

$$(3.2) \qquad \text{rk}(\mathcal{S}'_{I,i} \cap \mathcal{S}''_{J_0,q+1}/\mathcal{S}''_{J_0,q-1}) \mid Y_{w_0}$$
$$= \begin{cases} 0 \mid 0 & \text{for } i < a; \\ \delta_q(J_0) & \text{for } a \leq i < b; \\ \delta_q(J_0) + \delta_{q+1}(J_0) & \text{for } i \geq b. \end{cases}$$

$$(3.3) \qquad \text{rk}(\mathcal{S}'_{I,i} \cap \mathcal{S}''_{J,q}/\mathcal{S}''_{J,q-1}) \mid Y_w = \begin{cases} 0 \mid 0 & \text{for } i < b; \\ \delta_{q+1}(J_0) & \text{for } i \geq b. \end{cases}$$

(3.4) $\quad \operatorname{rk}(\mathcal{S}'_{I,i} \cap \mathcal{S}''_{J,q+1}/\mathcal{S}''_{J,q-1}) \mid_{Y_w} = \operatorname{rk}(\mathcal{S}'_{I,i} \cap \mathcal{S}_{J_0,q+1}/\mathcal{S}''_{J_0,q+1}) \mid_{Y_{w_0}}.$

(The last equality is valid for all i).

In fact, from Eq. (3.1), it follows that

$$\mathcal{S}''_{J_0,q}/\mathcal{S}''_{J_0,q-1} \mid_{Y_{w_0}} \subseteq \mathcal{S}'_{I,a} \cap \mathcal{S}''_{J_0,q+1}/\mathcal{S}''_{J_0,q-1} \mid_{Y_{w_0}},$$

and Eq. (3.2) shows that this is, in fact, an equality. Hence, the tautological sheaf $\mathcal{S}''_{J_0,q}/\mathcal{S}''_{J_0,q-1}$ of Y_w over Y coincides with the lift of $\mathcal{S}'_{I,a} \cap \mathcal{S}''_{J_0,q+1}/\mathcal{S}_{J_0,q-1}$ from Y so that $\dim Y_w = \dim Y$. Similarly, looking at Eqs. (3.3), (3.4), and (3.2), one concludes that the tautological sheaf $\mathcal{S}''_{J,q}/\mathcal{S}_{J,q-1}$ of rank $\delta_{q+1}(J_0)$ on Y_w over Y can be embedded in the sheaf $\mathcal{S}'_{I,b} \cap \mathcal{S}''_{J,q+1}/\mathcal{S}''_{J,q-1}$ of rank $\delta_q(J_0) + \delta_{q+1}(J_0)$ lifted from Y, and does not intersect there with the sheaf $\mathcal{S}'_{I,b-1} \cap \mathcal{S}''_{J,q+1}/\mathcal{S}''_{J,q-1}$ of rank $\delta_q(J_0)$ that is also lifted from Y. Thus,

$$\dim Y_w - \dim Y_{w_0} = \dim Y_w - \dim Y = l_{J_0 J}(\sigma^{k+1});$$

and finally,

$$\dim Y_w = l_{IJ}(w) + \dim F_I$$

by induction.

Case $\sigma^{k+1} = \tau_1$.

Consider the natural projections $F_{J_0} \to \overline{F}$ and $F_J \to \overline{F}$ onto a space of incomplete flags corresponding to deleting the lth component and its orthogonal complement in the case $\mathrm{OSp}(2r+1 \mid 2s)$ (only for these groups is this complement strictly larger). As above, one can show that the projections Y of Y_{w_0} and Y_w with respect to $\rho_0 : F_I \times F_{J_0} \to F_I \times \overline{F}$ and $\rho : F_I \times F_J \to F_I \times \overline{F}$ coincide and that $\dim Y_{w_0} = \dim Y$, while Y_w is a large cell of a relative grassmannian of isotropic lines. To be more precise, denote by $GI(k; l, b)$ a grassmannian of k-subspaces in an l-space isotropic with respect to a form b. Then the large cell mentioned above lies in $GI(\delta_l(J); \delta_l(J) + \delta_{l+1}(J) + \delta_{l+2}(J), b)$, if $G = \mathrm{OSp}(2r+1 \mid 2s)$, and in $GI(\delta_l(J); \delta_l(J) + \delta_{l+1}(J), b)$ otherwise.

In this way we get the following table:

Group	Grassmannian	Dimension
$\mathrm{OSp}(2r+1 \mid 2s)$	$GI(1 \mid 0; 3 \mid 0, b)$	$1 \mid 1$
$\mathrm{OSp}(2r+1 \mid 2s)$	$GI(0 \mid 1; 1 \mid 2, b)$	$1 \mid 1$
$\mathrm{OSp}(2r \mid 2s)$	$GI(1 \mid 0; 2 \mid 0, b)$	$0 \mid 0$
$\mathrm{OSp}(2r \mid 2s)$	$GI(0 \mid 1; 0 \mid 2, b)$	$1 \mid 0$
$\Pi\,\mathrm{Sp}(m)$	$GI(1 \mid 0; 1 \mid 1, b)$	$0 \mid 1$
$\Pi\,\mathrm{Sp}(m)$	$GI(0 \mid 1; 1 \mid 1, b)$	$0 \mid 0$

The last column of this table coincides with $l_{J_0J}(\sigma^{k+1})$ so that by inductive assumption, we get

$$\dim Y_w = l_{IJ}(w) + \dim F_I.$$

3.6. INCOMPLETE FLAGS. In the previous treatment, we exploited some properties of incomplete flags in order to be able to use induction. We shall now review the main features of incomplete flag spaces in a more systematic way.

We keep the notation of Section 1: ${}^G\mathbf{I}$ is the set of types of G-flags in the basic representation space $T^{m|n}$ of the group G; ${}^GF_I$ is the supermanifold of all flags of type I.

3.7. LEMMA. *We have*

$$\begin{aligned}
{}^{\mathrm{SL}}\mathbf{I} &= \{(\delta_1, \dots, \delta_r) \mid 1 \le r \le m+n, \delta_i > 0 \mid 0, \textstyle\sum \delta_i = m \mid n\};\\
{}^{\mathrm{OSp}}\mathbf{I} &= \{(\delta_1, \dots, \delta_r) \in {}^{\mathrm{SL}}\mathbf{I} \mid \delta_i = \delta_{r+1-i}\};\\
{}^{\Pi\,\mathrm{Sp}}\mathbf{I} &= \{(\delta_1, \dots, \delta_r) \in {}^{\mathrm{SL}}\mathbf{I} \mid \delta_i = \delta^c_{r+1-i}\};\\
{}^{Q}\mathbf{I} &= \{(\delta_1, \dots, \delta_r) \in {}^{\mathrm{SL}}\mathbf{I} \mid \delta_i = a_i \mid a_i\}.
\end{aligned}$$

With r fixed, consider a sequence of natural numbers $u = (u_1, \dots, u_r)$ such that $\sum u_i = m+n$. Denote by $\mathbf{D} = {}^G\mathbf{D}(r,u)$ the subset in ${}^G\mathbf{I}$ consisting of all types I with $|\delta_i(I)| = u_i$ for all i, where $|a \mid b| = a+b$. Put

$${}^GF_{\mathbf{D}} = \bigsqcup_{I \in \mathbf{D}} {}^GF_I.$$

3.8. Definition. Let $\mathcal{S}_I$ be a complete G-flag in T_S of type I, $\mathcal{S}_J$ a G-flag of type $I \in \mathbf{D}$. They are called *well-positioned* if all pairwise intersections of their components are locally free locally direct subsheaves of T_S of constant rank.

The type of relative position of these flags is the matrix with entries

$$d_{ij} = \mathrm{rk}(\mathcal{S}_{I,i} \cap \mathcal{S}_{J,j}), \quad 0 \le j \le r, \quad 0 \le i \le t,$$

where $t = m+n$ for $G = \mathrm{SL}, \mathrm{OSp}$; $t = 2m$ for $G = \Pi\,\mathrm{Sp}$; $t = m$ for $G = Q$ (this notation will be kept in the rest of this section).

3.9. LEMMA. *Any matrix of relative position of an incomplete G-flag with respect to a complete one can be obtained by deleting from a matrix of relative position of two complete G-flags $t-r$ rows with the numbers defined by the forgetting map ${}^G\mathbf{I}_c \to \mathbf{D}$.*

3.10. Definition. For $\mathbf{D} = {}^G\mathbf{D}(r,u)$ as above, we define

$$W_{\mathbf{D}} = \text{the subgroup of } {}^GW \text{ consisting of permutations that leave invariant all subsets } \{1,\dots,i_1\}, \{i_1+1,\dots,i_2\},\dots,\{i_r+1,\dots,t\}, \text{ where } i_k = \sum_{j=1}^{k} u_j.$$

We denote by $w(I)$ the image of the pair $w(I), w \in {}^GW/W_{\mathbf{D}}, I \in {}^G\mathbf{I}_c$ under the map ${}^GW/W_{\mathbf{D}} \times {}^G\mathbf{I}_c \to \mathbf{D}$ induced by the action ${}^GW \times {}^G\mathbf{I}_c \to {}^G\mathbf{I}_c$.

3.11. LEMMA. *There is a natural bijection between the following sets of data:*

(a) types of relative position of an incomplete G-flag of a type $I \in \mathbf{D}$ and a complete G-flag;

(b) matrices (d_{ij}) described in Lemma 3.9;

(c) triples $\{(I,J,w) \mid I \in {}^GI_c, j \in \mathbf{D}, w \in {}^GW/W_{\mathbf{D}}, J = w(I)\}$.

3.12. SCHUBERT SUPERCELLS. Put

$$F = {}^GF, \qquad F' = {}^GF_{\mathbf{D}}, \qquad W = {}^GW.$$

For each class $w \in W/W_{\mathbf{D}}$, denote by $d_{ij,w}$ the function on $\bigsqcup_{J=w(I)} F_I \times F'_J$ that on $F_I \times F'_J$ takes value $d_{ij,w,IJ}$ (a coefficient of the matrix corresponding to (I,J,w) in view of Lemma 3.11).

Define $|Y_w| \subset \bigsqcup_{J=w(I)} F_I \times F'_J$ as in the case of complete flags.

3.13. THEOREM. *(a) Every $|Y_w|$ can be endowed with a canonical structure of a locally closed subsuperscheme $Y_w \subset F \times F'$ in such a way that the union of Y_w is the flattening stratification for the family of sheaves $\mathcal{F}(\mathcal{S},\mathcal{S}')$.*

(b) Over F, all Y_w are relative affine superspaces.

(c) $\dim Y_w = \dim F = \min_{w' \in w \subset W} l(w'), \quad w \in W/W_{\mathbf{D}}$.

As in the case of complete flags, Schubert supercells are orbits of G. We leave the precise statement and proof to the reader.

4. Order in Flag Weyl Groups and Closure of Schubert Supercells

4.1. NOTATION. We return to the complete flags. For $I, J \in {}^G\mathbf{I}_c$, put $W_{IJ} = \{w \in {}^GW \mid w(I) = J\}, l(w) = l_0(w) \mid l_1(w) \in \mathbb{Z} \times \mathbb{Z}$ (the superlength of w). W_{JJ} are essentially the classical Weyl groups of G_{red}. The case $\text{OSp}(2r \mid 2s)$ is slightly anomalous, and we shall leave it to the reader to make the necessary changes in the statements and proofs. In that case, denote by W^0 the Weyl group of the identity component of G_{red}. It is a subgroup of index two in the corresponding Weyl group identified with W_{IJ}. In the definition below, for $G = \text{OSp}(2r \mid 2s)$, one should replace W_{IJ} by any of two orbits $W_{IJ}^{1,2}$ with respect to the action of $W^0 \subset W_{JJ}$ and W_{JJ} by W^0 itself.

4.2. Definition. A reflection in ${}^GW$ is an element conjugate to a basic reflection (cf. Section 3.1).

Let $w_1, w_2 \in W_{IJ}, \sigma \in W_{JJ}$ a reflection. We write $w_1 \xrightarrow{\sigma} w_2$, if $\sigma w_1 = w_2$ and $l_0(w_2) = l_0(w_1) + 1$.

Put $w < w'$ if there exists a chain $w = w_1 \to w_2 \to \cdots \to w_k = w'$. We consider this as a partial order on W_{IJ}.

Before stating the superanalog of the classical relation between closures of Schubert cells and the order in the Weyl groups, we recall some (super) scheme-theoretical notions.

Let $f : Y \to Z$ be a morphism. Its schematic image is a closed subsuperscheme $f(Y)$ in Z such that f can be decomposed via $f(Y) \subset Z$, and $f(Y)$ is universal with this property.

The schematic closure $\overline{Y}$ of a locally closed subscheme $Y \subset Z$ is the schematic image of the canonical embedding of Y.

If X and Y are locally closed subsuperschemes of Z, we shall write $X \subset \overline{Y}$ when the embedding $X \subset Z$ can be decomposed via $\overline{Y}$.

4.3. THEOREM. *Let $I, J \in {}^G\mathbf{I}_c, w, w' \in W_{IJ}$. Then the following conditions are equivalent.*

(a) $w \leq w'$.

(b) $d_{ij,w',IJ} \geq d_{ij,w,IJ}$ *(cf. Proposition 1.8).*

(c) $Y_{w',IJ} \subseteq \overline{Y_{w,IJ}}$.

(d) $(Y_{w',IJ})_{\text{red}} \subseteq \overline{(Y_{w,IJ})_{\text{red}}}$.

Proof. First of all, (a) $\Leftrightarrow$ (d) means that the order in W_{JJ} coincides with the standard order in the Weyl group of (the identity component of) the group G_{red}. For brevity, we shall omit mentioning connected components when necessary.

We shall prove a series of implications.

i. (a) $\Leftrightarrow$ (d).

4.4. LEMMA. *(1) There exists a unique element $w_0 \in W_{IJ}$ with $l_0(w_0) = 0$.*

(2) For all $w \in W_{JJ} = W$, we have $(Y_{ww_0})_{\text{red}} = {}^{G_{\text{red}}}Y_w$ with respect to the canonical identification $({}^GF)_{\text{red}} = {}^{G_{\text{red}}}F$.

(3) The order on W_{IJ} coincides with the standard order on W under the identification $W \to W_{IJ}, w \to ww_0$.

Proof of the Lemma. In the case $G = Q$, everything is obvious because $I = J = (1 \mid 1, \ldots, 1 \mid 1), w_0 = e, W_{JJ} = {}^G W$. From now on, we assume that $G \neq Q$.

(1) Reduced Schubert supercells of ${}^G F_I \times {}^G F_J$ constitute the system of Schubert cells for G_{red}. Among them, there is exactly one of dimension $0 + \dim({}^{G_{\text{red}}} F)$. Let it be the reduced space of Y_{w_0}, IJ. Theorem 3.3 shows that the even part of its dimension equals $l_0(w_0) + \dim({}^G F_{I,\text{red}})$. Hence, $l_0(w_0) = 0$.

(3) This statement follows from (2). In fact, if (2) is true then

$$\begin{aligned} l_0(ww_0) &= \dim(Y_{ww_0, IJ})_{\text{red}} - \dim({}^G F_I)_{\text{red}} \\ &= \dim({}^{G_{\text{red}}} Y_w) - \dim({}^{G_{\text{red}}} F) \\ &= N(w), \end{aligned}$$

where N(w) is the classical length of $w \in W = {}^{G_{\text{red}}} W$ with respect to the set of basic reflections associated with the Borel subgroup stabilizing the standard G-flag of the type J (cf. Lemma 1.8). Hence the order in W_{IJ} becomes the classical one by definition.

(2) Consider the map $W_{IJ} \longrightarrow {}^{G_{\text{red}}} W : w \longrightarrow w'$, where w' is well defined by the condition

$$(4.1) \qquad {}^{G_{\text{red}}} Y'_w = ({}^G Y_{w,IJ})_{\text{red}}.$$

In order to prove (2), we shall show that the same map can be defined by the right multiplication by a fixed element $\overline{w}$. Then from uniqueness and (1), it will follow that $\overline{w} = w_0^{-1}$.

Assume that Eq. (4.1) holds. Clearly, w' is determined by the type of relative position of a pair of G_{red}-flags $\varphi(f_I), f_J$ in $T^{m|n}$, where f_I, f_J are standard flags of respective types and φ is a certain transformation of f_I into a flag of type J. This transformation can be described as follows. Decompose each component $\mathcal{S}_i$ of the flag f_I into the direct sum of its even and odd parts: $\mathcal{S}_i = \mathcal{S}_{i0} \oplus \mathcal{S}_{i1}$. The flag $\varphi(f_I) = \mathcal{S}'$ is defined inductively. Assume that $\mathcal{S}'_i$ is already constructed in the form $\mathcal{S}_{i_0,0} \oplus \mathcal{S}_{i_1,1}$. If $\delta_{i+1}(J) = 1 \mid 0$, we put $\mathcal{S}'_{i+1} = \mathcal{S}_{k0} \oplus \mathcal{S}_{i_1,1}$, where k is the least integer satisfying $k > i_0, \delta_k(I) = 1 \mid 0$. If $\delta_{i+1}(J) = 0 \mid 1$, we put $\mathcal{S}'_{i+1} = \mathcal{S}_{i_0,0} \oplus \mathcal{S}_{k,1}$, where k is the least integer satisfying $k > i_1, \delta_k(I) = 0 \mid 1$. (For $G =$ OSp or ΠSp, one should consider only i, k with $i+1, k \leq [(m+n+1)/2]$ and then add the orthogonal components of the constructed half of the flag). In order to determine the permutation w' describing the relative position of $\varphi(f_I)$ and f_J, notice that in the ith column of the corresponding matrix, a jump occurs between rows

$w(i_0) - 1$ and $w(i_0)$ if $\delta_i(J) = 1 \mid 0$, and between rows $w(i_1) - 1$ and $w(i_1)$ otherwise. By definition of w', this means that

$$w'(i) = \begin{cases} w(i_0), & \text{if } \delta_i(J) = 1 \mid 0; \\ w(i_1), & \text{if } \delta_i(J) = 0 \mid 1. \end{cases}$$

Hence, $\overline{w} = (i_0, i)$ (resp. $\overline{w} = (i_1, i)$ verifies $w' = w\overline{w}$). This proves our lemma.

This reduces also the proof of the equivalence (a) $\Leftrightarrow$ (d) to a well-known classical fact.

ii. (d) $\Leftrightarrow$ (b).

This follows from the fact that the dimension of a geometric fiber of a coherent sheaf is an upper semicontinuous function of the point on the base.

iii. (b) $\Leftrightarrow$ (a).

Consider the set of the basic reflections in ${}^{G_{\text{red}}}W$ corresponding to the Borel subgroup in G_{red} stabilizing the standard G-flag of type J. Let Σ be the image of this set under the canonical isomorphism ${}^{G_{\text{red}}}W \to W_{JJ}$.

4.5. Lemma. *Assume that $d_{ij,w,IJ} \geq d_{ij,w',IJ}$ for all i, j. Then for any $\sigma \in \Sigma$, we have either $d_{ij,\sigma w,IJ} \geq d_{ij,w',IJ}$, or $d_{ij,\sigma w,IJ} \geq d_{ij,\sigma w',IJ}$ for all i, j.*

Proof of the Lemma. One can easily convince oneself that Σ consists of the following permutations. Denote by j_i the least integer j such that $j > i, \delta_j(J) = \delta_i(J)$. Then Σ is the family of (i, j_i) for SL; of $(i, j_i)(m + n + 1 - i, m + n + 1 - j_i)$ for OSp, ΠSp, completed by $(r + s, r + s + 2)$ for $\text{OSp}(2r + 1 \mid 2s)$ or $(r + s, r + s + 1)$ for $\text{OSp}(2r \mid 2s)$; or $(i, i + 1)$ for Q. For OSp, ΠSp, one should restrict oneself by $i, j_i \leq [(m + n + 1)/2]$.

The rest of the proof follows from the combinatorial properties of the relative position matrices, and we leave it to the reader.

Now we shall deduce from this lemma that (b) implies (a). Suppose that $d_{ij,w',IJ} \geq d_{ij,w,IJ}$ for all i, j. We shall prove that $w' \leq w$ by a downward induction by $l_0(w')$. From Lemma 4.4, (3), and the corresponding classical statement, it follows that if $w' = s \in W_{IJ}$ has the maximal value of $l_0(s)$, then $s \geq w$; hence, $d_{ij,s,IJ} \leq d_{ij,w,IJ}$ for all i, j (taking into account the implication (a) $\Rightarrow$ (b) already proved). This shows that if $d_{ij,s,IJ} \geq d_{ij,w,IJ}$ for all i, j, then, in fact, we have strict equalities, so that $s = w$, since a Schubert cell is determined by its relative position matrix.

Now suppose that $l_0(w') < l_0(s)$. Again from Lemma 4.4, (3), and a corresponding classical fact, it follows that $w' < \sigma w$ for an appropriate reflection $\sigma \in \Sigma$. By applying Lemma 4.5, we see that either $d_{ij,\sigma w',IJ} \geq d_{ij,w,IJ}$ or $d_{ij,\sigma w',IJ} \geq d_{ij,\sigma w,IJ}$ for all i, j. In the first case, the inductive assumption shows that $\sigma w' \leq w$, hence $w' \leq w$. In the second case, we similarly derive $\sigma w' \le \sigma w$ and then either $w' \le w$, or $\sigma w' \le w$, hence also $w' \le w$.

iv. (b) $\Rightarrow$ (c).

Since we have already proved that (a) and (b) are equivalent, it suffices to consider the case $w' = \sigma w, \sigma \in \Sigma \subset W_{JJ}, d_{ij,w,IJ} \ge d_{ij,w',IJ}$. We shall restrict ourselves to the case SL, $\sigma = (i_0, j_0), i_0 < j_0$, with the other cases being essentially similar.

Consider two projections identical on the first factor,

$$^{G}F_I \times {}^{G}F_I \xrightarrow{\psi} {}^{G}F_I \times F' \xrightarrow{\varphi} {}^{G}F_I \times F'',$$

and given on the second factors by the following prescriptions:

ψ : delete flag components $i_0 + 1, \dots, j_0 - 1$.

φ : delete flag component i_0.

Consider the relative position of the fibers of these projections and the cell intersection $Y_{w,IJ} \cap Y_{\sigma w,IJ}$. Notice first of all that the schematic images $\varphi\psi(Y_{w,IJ})$ and $\varphi\psi(Y_{\sigma w,IJ})$ coincide because these images are closures of Schubert cells for the space of incomplete flags F'', and the matrices determining these cells have the same columns with numbers $\le i_0$ and $\ge (j_0 - 1)$.

In the same manner, $\psi(Y_{w,IJ})$ and $\psi(Y_{\sigma w,IJ})$ are closures of Schubert supercells for the relative projective space $\mathrm{P}(\delta_{i_0}(J); \mathcal{S}_{i_0}/\mathcal{S}_{i_0-1})$ over $\varphi\psi(Y_{w,IJ}) = \varphi\psi(Y_{\sigma w,IJ})$. These cells are determined by the relative positions of $\mathcal{S}_{i_0}/\mathcal{S}_{i_0-1}$ and $\mathcal{S}_{I,k}/\mathcal{S}_{i_0-1}, 0 \le k \le m+n$, that, in turn, are determined by the matrices $(d_{ij,w,IJ})$ and $(d_{ij,\sigma w,IJ})$. The first matrix implies $\mathcal{S}_{i_0}/\mathcal{S}_{i_0-1} \subset (\mathcal{S}_{I,w^{-1}(i_0)} \cap \mathcal{S}_{j_0})/\mathcal{S}_{i_0-1}$ (complemented by a general position property), and the second matrix implies $\mathcal{S}i_0/\mathcal{S}_{i_0-1} \subset (\mathcal{S}_{I,w^{-1}(j_0)} \cap \mathcal{S}_{j_0})/\mathcal{S}_{i_0-1}$ complemented by a general position property.

From our assumption $d_{ij,w,IJ} \ge d_{ij,\sigma w,IJ}$ for all i, j, it follows that $w^{-1}(i_0) < w^{-1}(j_0)$ (cf. proof of Lemma 1.8). Hence, $\mathcal{S}_{I,w^{-1}(i_0)} \subset \mathcal{S}_{I,w^{-1}(j_0)}$ and

$$\psi(Y_{w,IJ}) \subset \psi(Y_{\sigma w,IJ}) = \overline{\psi(Y_{\sigma w,IJ})}.$$

Finally, $\overline{Y}_w$ and $\overline{Y}_{\sigma w}$ can be represented as intersections of $\psi^{-1}\psi(Y_w)$ and $\psi^{-1}\psi(Y_{\sigma w})$, respectively, with one and the same Schubert cell closure of the relative complete flag manifold ${}^{G}F_J \to F'$. This follows from the fact that the horizontal jumps in the submatrices consisting of columns $i_0, \dots, j_0 - 1$ are situated at the same places. Hence, we deduced that $\overline{Y}_w \subset \overline{Y}_{\sigma w}$ from $w \le \sigma w$.

v. (c) $\Rightarrow$ (d). Clear.

4.6. COROLLARY. *Suppose that* $I, J \in {}^{G}\mathbf{I}_c; w, w' \in W_{IJ}$. *Then* $l_0(w') \le l_0(w) \Leftrightarrow l(w') \le l(w)$.

5. Singularities of Schubert Supercells

5.1. SINGULARITIES OF SUPERSCHEMES. A point of a supervariety is called nonsingular if an open neighborhood of it is a supermanifold.

In this section, we shall study singularities and desingularization of Schubert varieties, that is, schematic closures of Schubert (super)cells.

In the pure even geometry, they play a significant role because the geometric picture of acquiring a singularity by specializing the position of a linear space is a universal one as long as linear approximation is sufficiently informative. There are reasons to believe that studying this class of singularities in supergeometry will also be able to shed some light on the general problem.

We shall describe here a superversion of the Bott–Samelson desingularization of Schubert varieties. The idea of this construction is a very intuitive one, and we shall illustrate it first by the simplest example. Assume that a variety Y parametrizes linear subspaces S of a fixed linear space T and acquires some singularities when $\dim(s \cap S_0)$ exceeds its stable value d (here S_0 is a fixed subspace). The idea is then to consider a new variety Y' parametrizing the pairs (S, S'), where $S' \subset S \cap S_0$ is a d-dimensional subspace. It clearly coincides with Y at an open subset, and the canonical morphism "forgetting S'": $Y' \to Y$ often is a desingularization of Y, or at least a partial one.

5.2. BOTT–SAMELSON SUPERSCHEMES. We shall keep the notation of the previous sections. In order to desingularize $\overline{Y}_{w,IJ}$, choose a reduced decomposition $w = s_k \dots s_1$ into a product of basic reflections. Put $w_i = s_i \dots s_1, 1 \leq i \leq k$. Define a tower of projectivized vector bundles ${}^G F_I = Z_0 \leftarrow Z_1 \leftarrow \cdots \leftarrow Z_k$ inductively. Suppose that Z_j is already constructed together with a flag $\mathcal{S} = \mathcal{S}^{(j)}$ of type $w_j(I)$ in $T_j = T \otimes \mathcal{O}_{Z_j}$, where, as always, T is the basic representation space of G. If $G = \mathrm{SL}$, we have $s_j = (i, i+1)$ for some i, and we put

$$Z_{j+1} = \mathbb{P}_{Z_j}(\delta_i(w_{j+1}(I)); \mathcal{S}_{i+1}/\mathcal{S}_{i-1}).$$

Its relative dimension over Z_j is

$${}^l w_j(I), w_{j+1}(I)({}^s j).$$

This space is equipped with the flag $\mathcal{S}^{(j+1)}$, whose components $\mathcal{S}_p^{(j+1)}$, $p \neq i$, are lifted from Z, while $\mathcal{S}_i^{(j+1)}$ is an extension of the relative tautological sheaf $\alpha^*(\mathcal{S}_i/\mathcal{S}_{i-1}) \subset \alpha^*(\mathcal{S}_{i+1}/\mathcal{S}_{i-1}), \alpha : Z_{j+1} \to Z_j$, by $\alpha^*(\mathcal{S}_{i-1})$, which is induced by the extension

$$0 \to \mathcal{S}_{i-1} \to \mathcal{S}_{i+1} \to \mathcal{S}_{i+1}/\mathcal{S}_{i-1} \to 0.$$

Consider now the cases $G = \mathrm{OSp}(m \mid n)$ and $\Pi\mathrm{Sp}(m)$. If $s_j = (i, i+1)(m+n+1-i, m+n-i), i < [(m+n)/2]$, we put again

$$Z_{j+1} = \mathbb{P}_{Z_j}(\delta_i(w_{j+1}(I)); \mathcal{S}_{i+1}/\mathcal{S}_{i-1}).$$

To define the structural flag in T_{j+1}, we lift all components from Z_j except of the ith and the $(m+n-i)$-th. The ith one is defined in the same way as for SL, and the $(m+n-1)$-th one as its orthogonal complement.

If $s_j = (l, m+n+1-l)$, $l = [(m+n)/2]$, put

$$Z_{j+1} = GI_{Z_j}(\delta_1(w_{j+1}(I)); \mathcal{S}_{m+n+1-l}/\mathcal{S}_{l-1}, b),$$

where GI means an isotropic grassmanian. The components of the structural flag are all lifted, with the exception of the lth and the $(m+n-l)$-th. The lth one is defined by means of the tautological sheaf of Z_{j+1}/Z_j, as above. For $m+n$ even, we obtain in this way a complete flag. For $m+n$ odd, we supplement it by the missing orthogonal complement.

Finally, the case $G = Q(m)$ is similar to $\mathrm{SL}(m \mid 0)$. By construction, all Z_j are supermanifolds.

The last scheme $Z_w = Z_k$ is called a Bott–Samelson scheme (in fact, it depends not only on w but also on the reduced decomposition of w), but for brevity we shall still denote it by Z_w.

5.3. Bott–Samelson Morphism. Let $\mathcal{S}$ be the structural flag of the type $J = w(I)$ on Z_w, and $\beta : Z_w \to Z_0 = {}^G F_I$ the natural projection. We define the Bott–Samelson morphism

$$\psi : Z_w \to {}^G F_I \times {}^G F_J$$

by the conditions

$$\psi^*(\mathcal{S}_I) = \beta^*(\mathcal{S}_I), \qquad \psi^*(\mathcal{S}_J) = \mathcal{S}.$$

This is a morphism of superschemes over ${}^G F_I$.

5.4. Theorem. *The schematic image of the Bott–Samelson morphism is* $\overline{Y}_{w,IJ}$. *It induces an isomorphism of an open subscheme of* Z_w *with* $Y_{w,IJ}$.

Proof. We start by describing the open submanifold $U_w \subset Z_w$ that is mapped isomorphically onto $Y_{w,IJ}$. It will be the last floor of a tower $U_j \subset Z_j$.

Denote by $\psi_j : Z_j \to {}^G F_I \times {}^G F_{w_j(I)}$ a morphism defined by properties similar to those of ψ. Put $U_0 = Z_0 = {}^G F_I$. Clearly, ψ_0 defines an isomorphism of U_0 with the diagonal of ${}^G F_I \times {}^G F_I$, that is, with $Y_{\mathrm{id},II}$.

If U_j is already constructed and we know that

$$\psi_j \mid_{U_j} : U_j \to Y_{w_j, I, w_j(I)}$$

is an isomorphism, we take for U_{j+1} a large cell in the relative grassmannian

$$Z'_{j+1} = \beta_j^{-1}(U_j) \to U_j,$$

where $\beta_j : Z_{j+1} \to Z_j$ is the natural projection. The cell decomposition we have in mind is induced by the flag of type I, lifted from ${}^G F_I$ with the help of the morphism $Z'_{j+1} \to U_j \to U_0 = {}^G F_I$. In order to see that ψ_{j+1} is an isomorphism, notice that it is induced by an identification of tautological sheaves and that ψ_j is an isomorphism by the inductive assumption.

Now that we know that $U_w \to Y_{w,IJ}$, we can conclude that ψ defines a morphism $\overline{U}_w \to \overline{Y}_{w,IJ}$ such that the Schubert variety is the schematic image of $\overline{U}_w = Z_w$. This concludes the proof.

5.5. AN EXAMPLE. Let us make explicit calculations for a Schubert cell in a grassmannian $X = G(2 \mid 0; T), T = \mathbb{C}^{3|1}$. For simplicity, we shall fix a standard flag instead of letting it vary in a complete flag space.

We have $\dim(X) = 2 \mid 2, X_{\text{red}} = \mathbb{P}^2$. Denote by $p : X \to \mathbb{P}(1 \mid 0; \Lambda^2(T))$ the Plücker embedding into $\mathbb{P}^{3|3}$. In terms of points, the image of p consists of even decomposable bivectors.

Let $\{e_1, e_2, e_3, f\}$ be a basis in T (e_i even, f odd). Consider a bivector $Q = Q_1 + Q_2 \wedge f + \lambda f \wedge f$, where

$$Q_1 = \lambda_{12} e_1 \wedge e_2 + \lambda_{13} e_1 \wedge e_3 + \lambda_{23} e_2 \wedge e_3, \quad \lambda_{ij} \text{ and } \lambda \text{ even };$$
$$Q_2 = \lambda_1 e_1 + \lambda_2 e_2 + \lambda_3 e_3, \quad \lambda_i \text{ odd}.$$

Decomposability of Q means that

$Q = (R + af) \wedge (S + bf) = R \wedge S + (bR - aS) \wedge f - abf \wedge f$, where R, S do not depend on f; a, b are odd. An equality $Q_1 = R \Lambda S$ puts no restrictions upon Q_1. If we fix R and S, then the equality $Q_2 = bR - aS$ for some a, b means that $Q_1 \Lambda Q_2 = 0$, while $\lambda = -ab$ means that $Q_2 \Lambda Q_2 + 2\lambda Q_1 = 0$. This finally gives equations of the Plücker image of our grassmannian:

$$\lambda_1 \lambda_{23} - \lambda_2 \lambda_{13} + \lambda_3 \lambda_{12} = 0;$$
$$\lambda_1 \lambda_2 + \lambda \lambda_{12} = 0;$$
$$\lambda_1 \lambda_3 + \lambda \lambda_{13} = 0;$$
$$\lambda_2 \lambda_3 + \lambda \lambda_{23} = 0.$$

Now consider the Schubert cell parametrizing the subspaces of T that are well-positioned with $V = \langle e_1, e_3 \rangle$ and have a one-dimensional intersection with this subspace. Its closure is given by the equations

$$\lambda_1 \lambda_{23} + \lambda_3 \lambda_{12} = \lambda_1 \lambda_3 = \lambda_2 = \lambda = 0.$$

This supervariety is nonsingular outside of the point

$$\lambda_{13} = 1; \quad \text{the rest of coordinates} = 0$$

of the reduced space.

5.6. THE BOTT–SAMELSON RESOLUTION. In this particular situation, put $M = \mathbb{P}(1 \mid 0; V) \times \mathbb{P}(1 \mid 0; T/V)$, $Z = \mathbb{P}_M(1 \mid 0; \mathcal{S}_3/\mathcal{S}_1)$, where $\mathcal{S}_1$ is the tautological sheaf lifted from the first factor, and $\mathcal{S}_3$ is the canonical extension of the tautological sheaf lifted from the second factor by $V \times \mathcal{O}_Z$. The Bott–Samelson morphism $\psi : Z \to G(2 \mid 0; T)$ is defined by the condition that the inverse image of the tautological sheaf is $\mathcal{S}_2$, where $\mathcal{S}_1 \subset \mathcal{S}_2 \subset \mathcal{S}_3$ and $\mathcal{S}_2/\mathcal{S}_1$ is the tautological sheaf on $Z = \mathbb{P}_M$.

In order to describe ψ in coordinates, consider a neighborhood U of the singular point $s \in \overline{Y}$ in $\mathbb{P}^{3|3}$. By dividing λ_{ij}, λ_k by λ_{13}, we get the affine coordinates l_{ij}, l_k in U. The homogeneous coordinates on the two factors of M will be denoted by (a_1, a_2) and (b, β), respectively. We shall write down the equations of an open submanifold of Z embedded into $U \times \mathbb{P}(1 \mid 0; V) \times \mathbb{P}(1 \mid 0; T/V)$. They consist of equations of $X = G(2 \mid 0; T)$ and incidence equations $\mathcal{S}_1 \subset \mathcal{S}_2 \subset \mathcal{S}_3$. Explicitly, let $\mathcal{S}_1 = \langle a_1 e_1 + a_2 e_2\rangle, \mathcal{S}_3 = \langle bf^* + \beta e_2\rangle^{\perp} \subset T$ (the asterisk refers to the dual basis in T^*), and let $\mathcal{S}_2$ correspond to the bivector

$$Q = l_{12} e_1 \wedge e_2 + e_1 \wedge e_3 + \lambda_{23} e_2 \wedge e_3 + l_1 e_1 \wedge f \\ + l_2 e_2 \wedge f + l_3 e_3 \wedge f + l f \wedge f \in \Lambda^2(T)_0.$$

Then $\mathcal{S}_1 \subset \mathcal{S}_2$ means that $(a_1 e_1 + a_2 e_3)\Lambda Q = 0$ and $\mathcal{S}_2 \subset \mathcal{S}_3$ is equivalent to $i(bf^* + \beta e_2^*)Q = 0$, where $i(\,.\,)$. denotes the inner product. This leads to the equations

$$a_1 l_{23} + a_2 l_{12} = a_1 l_3 - a_2 l_1 = a_1 l_2 = a_2 l_2 = a_1 l = a_2 l = 0, \\ b l_1 - \beta l_{12} = b l_3 + \beta l_{23} = 2bl + \beta l_2 = b l_2 = 0.$$

Putting everything together, we find the following equations of Z in $U \times \mathbb{P}^{1|0} \times \mathbb{P}^{0|1}$:

$$a_1 l_{23} + a_2 l_{12} = b l_1 - \beta l_{12} = b l_3 + \beta l_{23} = l = l_2 = 0.$$

One sees that this part of Z is nonsingular.

The Bott-Samelson morphism consists of forgetting a_1, a_2, b, β. Hence, the preimage of the singular point of $\overline{Y}_{\text{red}}$ is precisely $\mathbb{P}^{1|0} \times \mathbb{P}^{0|1}$.

6. Root Systems and Parabolic Subgroups

6.1. FLAG STABILIZERS AND PARABOLIC SUBGROUPS. In classical geometry,one can define a parabolic subgroup P of a simple complex algebraic group G by one of the following equivalent conditions:

(i) P contains a Borel subgroup B (all of which are conjugate).
(ii) P is the connected component of a stabilizer of a G-flag (for classical series).
(iii) G/P is proper.

In supergeometry the third definition is broader; in fact, it is equivalent to the condition that P_{red} is a parabolic subgroup of G_{red}.

Here we shall reproduce a result of Voronov that definitions (i) and (ii) are still equivalent for SL, OSp, and Q if a Borel subgroup *is defined* in supergeometry as a stabilizer of a complete flag. (However, not all Borel subgroups are conjugate. Their conjugacy classes are enumerated by the types of complete flags.)

One of the reasons for including this section is our desire to illustrate some peculiarities of the root technique in supergeometry stemming first of all from the fact that most of the appearing root systems do not satisfy the standard definition of an abstract root system.

For further reading, we suggest Penkov's papers [Pe], [PS] and his forthcoming book on the super-Borel–Weyl–Bott theory, where this subject is treated much more thoroughly.

6.2. BOREL AND PARABOLIC SUBGROUPS. For $G = \mathrm{SL}, \mathrm{OSp}, \Pi\,\mathrm{Sp}$, or Q, we shall call a *Borel subgroup* the stabilizer of a complete G-flag in the basic representation space T. In all cases except for Q, such a stabilizer consists of upper triangular matrices whose format is determined by the flag type.

For $G = Q \subset \mathrm{GL}(T)$ and the standard format realization of $\mathrm{GL}(T)$, G consists of four upper triangular blocks.

A *parabolic subgroup* is, by definition, a closed connected subgroup containing a Borel subgroup.

6.3. REVIEW OF ROOT SYSTEMS. Let $\mathfrak{g}$ be a classical simple Lie superalgebra belonging in Kac's notation [K] to one of the following types:

$$A(m \mid n), m \neq n, m, n \geq 0; \quad B(m \mid n), m \geq 0, n \geq 0;$$
$$C(n), n \geq 2; D(m \mid n), m \geq 2, n \geq 0; \quad D(2 \mid 1; \alpha), F(4) \text{ or } G(3).$$

In our notation,

$$A(m \mid n) = \mathrm{sl}(m+1 \mid n+1); \quad B(m \mid n) = \mathrm{osp}(2m+1 \mid 2n);$$
$$C(n) = \mathrm{osp}(2 \mid 2n-2); \quad D(m \mid n) = \mathrm{osp}(2m \mid 2n).$$

We use the notion and the properties of a Cartan subalgebra studied in [K1]. We shall treat $q(n)$ separately.

A Cartan subalgebra $\mathfrak{h} \subset \mathfrak{g}$ determines the decomposition of $\mathfrak{g}$ into root subspaces

$$\mathfrak{g} = \oplus_{\alpha \in \Delta} \mathfrak{g}_\alpha$$

with the following properties.

6.4. Proposition. *(Cf. [K1], 2.5.5)*

(a) $\mathfrak{g}_0 = \mathfrak{h}$.

(b) $\dim(\mathfrak{g}_\alpha) = 1$ *for all* $\alpha \neq 0$; $\alpha \in \Delta \Leftrightarrow -\alpha \in \Delta$.

(c) There is one and only one, up to a scalar, nondegenerate even symmetric bilinear form (,) on $\mathfrak{g}$.

(d) $[\mathfrak{g}_\alpha, \mathfrak{g}_\beta] = \mathfrak{g}_{\alpha+\beta} \Leftrightarrow \alpha, \beta, \alpha + \beta \in \Delta, \alpha + \beta \neq 0$.

(e) $(\mathfrak{g}_\alpha, \mathfrak{g}_\beta) = 0$ *if* $\alpha \neq -\beta$.

(f) Pairing (,) between $\mathfrak{g}_\alpha$ *and* $\mathfrak{g}_\alpha$ *is nondegenerate.*

(g) $[e_\alpha, e_{-\alpha}] = \alpha(h)h_\alpha$, *where* $h_\alpha \neq 0$ *is determined by* $(h_\alpha, h) = \alpha(h)$ *for all* $h \in \mathfrak{h} : \mathfrak{e}_{\pm\alpha} \in \mathfrak{g}_{\pm\alpha}$.

6.5. Definition. (Cf. [K1], 2.5.4) A subset $\Pi = \{\alpha_1, \ldots, \alpha_r\}$ is called a *system of simple roots*, if it is a minimal set with the following properties: There exists a system of generators $\{e_i, f_i, h_i\}$ such that $e_i \in \mathfrak{g}_{\alpha_i}, \mathfrak{f}_i \in \mathfrak{g}_{-\alpha_i}, [\mathfrak{e}_i, \mathfrak{f}_j] = \delta_{ij}\mathfrak{h}_j \in \mathfrak{h}; \mathfrak{i} = 1, \ldots, \mathfrak{r}$.

For all classical Lie superalgebras except $A(m \mid n)$, the system $\Delta \backslash \{0\}$ is not an abstract root system. Nevertheless, a number of properties of abstract root systems necessary for the proof of the main result of this section is still available.

6.6. Proposition. *There exists a system of simple roots* $\Pi \subset \Delta$. *Any such system has the following properties.*

(a) Π *is a base of* $\mathfrak{h}^*$.

(b) Any root $\beta \in \Delta$ *is a linear combination of simple roots with integer coefficients of one and the same sign.*

Proof of (b). Let $\mathfrak{g}_+$ (resp. $\mathfrak{g}_-$) be the subalgebra of $\mathfrak{g}$ generated by $\{e_i\}$ (resp. $\{f_i\}$). From Proposition 6.4 it follows that $\mathfrak{g} = \mathfrak{g}_- \oplus \mathfrak{h} \oplus \mathfrak{g}_+$. A multiple commutator of e_i's (resp. f_i's) belongs to $\mathfrak{g}_\beta$, where $\beta = \Sigma m_i \alpha_i, m_i$ (resp. $-m_i$) is the degree of this commutator with respect to e_i (resp. f_i).

Proof of (a). The existence of Π is proved and all Π's are explicitly described in [K1], 2.5.4. Looking at the list, one sees that Π is linearly independent.

Moreover, Π generates $\mathfrak{h}^*$ as a linear space. In fact, for $\alpha \in \mathfrak{h}^*$, consider $h_\alpha \in \mathfrak{h}$ such that $(h_\alpha, h) = \alpha(h)$ for all $h \in \mathfrak{h}$. Since $\mathfrak{g}$ is generated by e_i, f_j, it follows that h_α is a linear combination of $h_i = [e_i, f_i] = (e_i, f_i)h_{\alpha_i}$. Hence, α is a linear combination of α_i.

POSITIVE AND INDECOMPOSABLE ROOTS. If we put

$$\Delta^+ = (\Delta \cap \mathbb{Z}_+\Pi)\backslash\{0\}, \qquad \Delta^- = (\Delta \cap \mathbb{Z}_-\Pi)\backslash\{0\},$$

we have $\Delta = \Delta^+ \cup \Delta^- \cup \{0\}, \Delta^+ \cap \Delta^- = \emptyset$. As in the classical case, a positive root $\alpha \in \Delta^+$ is called *decomposable* if it is a sum of two positive roots. Also as in the classical case, the set of indecomposable roots coincides with Π.

A choice of a positive root system in a classical Lie superalgebra $\mathfrak{g}$ defines the corresponding Borel subalgebra

$$\mathfrak{b} = \oplus_{\alpha \in \Delta^+ \cup \{0\}} \mathfrak{g}_\alpha.$$

Any subalgebra containing a Borel one will be called *parabolic*.

6.7. PROPOSITION. *Let G be a complex algebraic supergroup of the type* $\mathrm{SL}(m \mid n), m \neq n$, *or* $\mathrm{OSp}(m \mid n)$, *and $\mathfrak{g}$ its Lie superalgebra. Then the standard bijective correspondence between algebraic Lie subsuperalgebras in $\mathfrak{g}$ and closed connected subgroups in G induces a bijective correspondence between Borel (resp. parabolic) subalgebras and subgroups.*

Proof. Let $\mathfrak{b}$ be a Borel subalgebra in $\mathfrak{g}$, corresponding to a positive root system Δ^+. Each irreducible finite-dimensional representation of $\mathfrak{b}$ has dimension $1 \mid 0$ or $0 \mid 1$, because a highest-weight vector is a proper vector. It follows that the basic representation of $\mathfrak{b}$ on T leaves invariant a complete SL-flag f. If $G = \mathrm{SL}$, this means that the closed subgroup B corresponding to $\mathfrak{b}$ is contained in the stabilizer of this flag $S_G(f)$ that is a Borel subgroup. In fact, $B = S_G(f)$. Otherwise, among the roots of the Lie superalgebra of $S_G(f)$, we could find a pair $\{\alpha, -\alpha\}$. But one can directly check that this is impossible.

If $G = \mathrm{OSp}$, one should argue slightly more carefully. First notice that any weight vector in the basic representation of $\mathfrak{g}$ is isotropic (the same is true for any subquotient of the restriction of this representation to $\mathfrak{b}$). In fact, all weights of the basic representation are nonzero, and if $v \in T$ is of weight α, we have

$$0 = \langle hv, v\rangle + \langle v, hv\rangle = 2\alpha(h)\langle v, v\rangle,$$

where $\langle\,,\rangle$ is the invariant bilinear form on T, so that $\langle v, v\rangle = 0$. Now we shall find by induction a complete $\mathfrak{b}$-invariant G-flag, starting with an isotropic invariant-weight subspace $V \subset T$, constructing a new representation space $V^{\perp}/V$, and repeating this procedure. As for $G = \mathrm{SL}$, one can check that B coincides with the stabilizer of this flag.

Now start with a Borel subgroup B in G. As we already mentioned, if α is a root of its Lie superalgebra $\mathfrak{b}$, then $-\alpha$ is not one. Denote by Δ^+ the system of nonzero roots of $\mathfrak{b}$, by Δ^- its complement in $\Delta\backslash\{0\}$. We have $\Delta^- = -\Delta^+$. Let $\Pi = \{\alpha_1, \dots, \alpha_r\}$ be the set of indecomposable elements of Δ^+. We shall show that Π is a system of simple roots in the sense of Section 6.5.

Choose nonzero vectors $e_i \in \mathfrak{g}_{\alpha_i}, \mathfrak{f}_i \in \mathfrak{g}_{-\alpha_i}$ and put $h_i = [e_i, f_i]$. We have $h_i \in \mathfrak{h}$. Besides, $[e_i, f_j] = 0$ for $i \neq j$, because otherwise $\alpha_i - \alpha_j = \beta \in \Delta\backslash\{0\}$, and this contradicts the indecomposability of α_i and α_j.

Clearly, the $\{e_i\}$ generate the subalgebra $\oplus_{\alpha\in\Delta^+}\mathfrak{g}_\alpha$ and similarly for $\{f_i\}$. Furthermore, $\{h_i\}$ linearly generate $\mathfrak{h}$, because Π generates $\mathfrak{h}^*$, and the $(\,,\,)$-dual basis of $\{h_i\}$ consists of functionals proportional to the roots α_i. It follows that the $\{e_i, f_j\}$ generate $\mathfrak{g}$. Finally, Π is minimal among subsets with these properties, because any subset $\Pi' \subseteq \Pi$ that is a simple root system must define a Borel subalgebra $\mathfrak{b}'$ contained in $\mathfrak{b}$. But, as we have shown in the first part of the proof, $\mathfrak{b}' = \mathfrak{b}$, so that $\Pi' = \Pi$ and $\mathfrak{b}$ is a Borel subalgebra.

6.8. THEOREM. *Let $\mathfrak{g}$ be a classical Lie superalgebra of one of the types listed in Section 6.3, $\mathfrak{b}$ a Borel subalgebra, Π the corresponding simple root system. Then for any subalgebra $\mathfrak{p} \supseteq \mathfrak{b}$, there exists a subset $I \subseteq \Pi$ such that*

$$\mathfrak{p} = \mathfrak{b} \oplus (\oplus_{\alpha\in\Delta_I^-}\mathfrak{g}_\alpha),$$

where Δ_I^- is the intersection of Δ^- with the semigroup generated by I. This correspondence is bijective.

Proof. Since $\mathfrak{h}$ is an Abelian subalgebra of $\mathfrak{p}$, we have $\mathfrak{p} = \oplus_{\alpha\in A}\mathfrak{g}_\alpha$ for a subset $A \subseteq \Delta$. We want to prove that $A = \Delta^+ \cup \{0\} \cup \Delta_I^-$ for some $i \subseteq \Pi$.

For I, take the set of all simple roots appearing in the decomposition of roots in $A \cap \Delta^-$. Let us prove that $I \subset A$. Suppose that $\alpha \in A \cap \Delta^-$ and α is decomposable, i.e., $-\alpha = \beta + \gamma$ for certain $\beta, \gamma \in \Delta^+$. Then $-\beta, -\gamma \in A \cap \Delta^-$, because $[\mathfrak{g}_\alpha, \mathfrak{g}_\beta] = \mathfrak{g}_{-\gamma}$ and $[\mathfrak{g}_\alpha, \mathfrak{g}_\gamma] = \mathfrak{g}_{-\beta}$. Since $\beta, \gamma \in \Delta^+ \subset A$, we have $-\beta, -\gamma \in A$. By inductive reasoning, we get $I \subset A$ and then $\Delta_I^- \subset A$. Since $A \cap \Delta^- \subset \Delta_I^-$ by definition of I, we finally get $A = \Delta^+ \cup \{0\} \cup \Delta_I^-$.

6.9. COROLLARY (I. SKORNYAKOV). *All connected parabolic subgroups in G are stabilizators of G-flags in T (for $G = \mathrm{SL}$ or OSp).*

Proof. Let B be a connected parabolic subgroup stabilizing a complete G-flag $f = \{0 \subset S_1 \subset S_2 \subset \cdots \subset S_{m+n} = T\}$. We shall treat the two situations separately.

(i) $G + \mathrm{SL}(m \mid n), \mathrm{OSp}(2r+1 \mid 2s)$ *or* $\mathrm{OSp}(2r \mid 2s)$;*and in the latter case* $\dim S_{r+s}/S_{r+s-1} = 0 \mid 1$.

Let f' be a (possibly noncomplete) flag all of whose components are components of f. Clearly, $S_G(f') \supset S_G(f) = B$ (here S_G means the identity component of a stabilizer). Hence, $S(f')$ is a parabolic subgroup. Therefore, the set of stabilizers of subflags of f consists of some parabolic subgroups containing B. Let us now show that both sets have the same cardinality. We know that the number of parabolic subgroups containing B equals the number of subsets of the corresponding simple root system Π. We have

$$G = \mathrm{SL} : \mathrm{card}(\Pi) = m + n - 1;$$
$$G = \mathrm{OSp} : \mathrm{card}(\Pi) = [(m+n)/2].$$

But this number coincides with the number of components (isotropic ones for OSp) of f, not counting 0 and T. In turn, the number of subsets of this last set coincides wit the number of G-subflags of f.

Different subflags f_1', f_2' have different stabilizers. Otherwise the flag f' consisting of all components of f_i' would have the same stabilizer as its proper subflag. But this is impossible in the situation considered: One can check directly that if a component of f' is missing in f_1', it can be deformed (at least infinitesimally) in a family of G-flags. This shows that there are as many stabilizers of subflags of a complete G-flag f as of connected parabolic subgroups containing the stabilizer of f.

(ii) $G = \mathrm{OSp}(2r \mid 2s)$ *and* $\dim S_{r+s}/S_{r+s-1} = 1 \mid 0$.

In this case, the Borel subgroup B stabilizing f is simultaneously stabilizing an incomplete flag $f_1' = \{f$ *without* $S_{r+s}\}$. Moreover, f_1' can be completed in exactly one more way to a G-flag f', whose stabilizer is still B. The inherent reason lies in the existence of two families of Lagrangian subspaces of T. Let S_{r+s}' be the corresponding component of f'.

As in the case (i), we notice first that the set of stabilizers of subflags of f and f' is a part of the set of parabolic subgroups containing B. Now we shall calculate their cardinalities. The second one consists of $2^{|\Pi|} = 2^{r+s}$ elements. The first one is the disjoint union of the stabilizers of the following G-flags:

(1) Flags containing S_{r+s} but not S_{r+s-1}.
(2) Flags containing S_{r+s}' but not S_{r+s-1}.
(3) Flags containing both S_{r+s} and S_{r+s-1}.
(4) Flags that do not contain S_{r+s}, S_{r+s}' and S_{r+s-1}.

Of course, components of these flags are to be chosen from those of f and f'. One sees that each family consists of 2^{r+s-2} elements.

6.10. CASE $G = Q$. We shall briefly explain the necessary modifications in the statements and proofs.

Let $\mathfrak{g} = \mathfrak{q}(\mathfrak{n}) = \{\mathfrak{X} \in \mathrm{gl}(\mathfrak{n} \mid \mathfrak{n}) \mid [\mathfrak{X}, \mathfrak{p}] = 0\}$, and let p be a fixed odd involution in T. Let $\mathfrak{h}_0$ be a Cartan subalgebra of the even part $\mathfrak{g}_0 = \mathrm{gl}(\mathfrak{n})$. It defines the root space decomposition

$$\mathfrak{g} = \oplus_{\alpha\in\Delta}\mathfrak{g}_\alpha$$

with the following properties (cf. [Pe]):

(a) Δ is the root system of $\mathfrak{g}_0$ (hence A_{n-1}).

(b) $\dim \mathfrak{g}_\alpha = 1 \mid 1$ for all $\alpha = 0$.

(c) $[\mathfrak{g}_\alpha, \mathfrak{g}_\beta] = \mathfrak{g}_{\alpha+\beta} \Leftrightarrow \alpha, \beta, \alpha+\beta \in \Delta, \alpha+\beta \neq 0$.

In this case, Δ is an abstract (complex) root system in the subspace of $\mathfrak{h}_0^*$ vanishing on the center of $\mathrm{gl}(n)$. We shall call a subset $\Pi = \{\alpha_1, \ldots, \alpha_r\} \subset \Delta$ a simple root system if it is an abstract simple root system. A choice of Π determines Δ^+ and Δ^- in the usual way, and we can define a Borel subalgebra

$$\mathfrak{b} = \oplus_{\alpha\in\Delta^+\cup\{0\}}\mathfrak{g}_\alpha.$$

A (standard) parabolic subalgebra is a subalgebra containing $\mathfrak{b}$. Enumeration of the standard parabolic subgroups in our sense is equivalent to that of the standard parabolic subalgebras.

6.11. THEOREM. *For $\mathfrak{g} = \mathfrak{q}(\mathfrak{n}), \mathfrak{n} \geq 3$, the standard parabolic subalgebras are in one-to-one correspondence with subsets of* Π, *and parabolic subgroups are stabilizers of G-flags.*

This is also true for $q(1), q(2)$, and $sq(n)$ for $n \geq 3$, where $sq(n)$ consists of endomorphisms with vanishing *odd trace*. However, $sq(2)$ contains a parabolic subalgebra that is not a flag stabilizer.

A description of Borel subalgebras of $\Pi\,\mathrm{Sp}$ is given in [Pe]. From the classical viewpoint, the root system of $\Pi\,\mathrm{Sp}$ is very peculiar.

CHAPTER 4

Quantum Groups as Symmetries of Quantum Spaces

1. Quantum Supergroups

1.1. HOPF SUPERALGEBRAS. In this chapter, we shall introduce and investigate a method of construction of quantum groups considered as "automorphisms objects" of noncommutative spaces (we continue to act in the framework of noncommutative *algebraic* geometry). A part of our results will be stated for quantum supergroups and superspaces, represented by $\mathbb{Z}_2$-graded algebras and grade-preserving morphisms. This means that the definitions and results of Chapter 1, Section 3 should be slightly refined.

We shall review here the main differences. Recall that the compatibility axiom of a bialgebra looks formally the same (see Chapter 1, Section 3.2(c)), but the multiplication in $E \otimes E$ is now given by the formula $(a \otimes b)(c \otimes d) = (-1)^{\hat{b}\hat{c}} ac \otimes bd$, where as usual, $\hat{b}$ is the parity of b. Similar signs enter in the definition of m^{op} and Δ^{op} (of course, this is true if we want to use elements; from the viewpoint of tensor categories explained in Chapter 1, Section 3, nothing changes except for the definition of $S_{(12)}$ in the basic tensor category).

A slightly less obvious addition should be made in the discussion of multiplicative matrices.

We shall call a *format* of a square matrix any sequence $(a_1, \dots, a_n)$, $a_i \in \mathbb{Z}_2$. A matrix $Z = (z_i^k)$ with coefficients in a $\mathbb{Z}_2$-graded space has this format if $\hat{z}_i^k = a_i + a_k$ (then it also has the format $(1 - a_1, \dots, 1 - a_n)$). Putting $\hat{\imath} = a_i$, we define the supertransposed matrix of this format by

$$(Z^{\mathrm{st}})_i^k = (-1)^{\hat{k}(\hat{\imath}+1)} (Z)_k^i.$$

Note that if we replace the format of Z by $(1 - \hat{\imath})$, Z^{st} will be changed. Therefore one should rather write $(Z, \mathit{format})^{\mathrm{st}}$. However, we shall omit format in the notation and prescribe to Z^{st} the same format as to Z. We have $(\mathrm{st})^4 = \mathrm{id}$ but $(\mathrm{st})^2 \neq id$ in general.

A *multiplicative matrix* is a matrix Z with coefficients in a coalgebra (E, Δ, ε), *endowed with a format*, and such that $\Delta(Z) = Z \otimes Z$, $\varepsilon(Z) = I$.

An analog of Proposition 3.5 in Chapter 1 is valid in this situation and explains why one needs the notion of a format: Z acts upon the superspace $\otimes k e_i$ such that $\hat{e}_i = \hat{\imath}$.

We leave to the reader the calculation of the Kronecker product of two multiplicative matrices.

1.2. LEMMA. *(a) Let Z be a matrix with entries in a supercoalgebra (E, Δ, ε) endowed with a format. Then*

$$(1.1)\quad \Delta(Z) = Z \otimes Z \Leftrightarrow \Delta^{\mathrm{op}}(Z^{\mathrm{st}}) = Z^{\mathrm{st}} \otimes Z^{\mathrm{st}}$$

(b) Let Z be a multiplicative matrix with coefficients in a Hopf superalgebra. Put $Z_k = i^k(Z)$. Then $Z_k^{\mathrm{st}^k}$ is multiplicative and

$$(1.2)\quad Z_k^{\mathrm{st}^k} Z_{k+1}^{\mathrm{st}^k} = Z_{k+1}^{\mathrm{st}^k} Z_k^{\mathrm{st}^k} = I.$$

Proof. (a) Let $Z = (z_\imath^k)$, $\Delta(Z) Z \otimes Z$. Then

$$\begin{aligned}
(\Delta^{\mathrm{op}}(Z^{\mathrm{st}}))_i^k &= (-1)^{\hat{k}(\hat{\imath}+1)} \Delta^{\mathrm{op}}(Z)_k^i = (-1)^{\hat{k}(\hat{\imath}+1)} S_{12}\Big(\sum_j z_k^j \otimes z_j^\imath\Big) \\
&= (-1)^{\hat{k}(\hat{\imath}+1)} \sum_j (-1)^{(\hat{k}+\hat{\jmath})(\hat{\imath}+\hat{\jmath})} z_j^i \otimes z_k^j, \\
(Z^{\mathrm{st}} \otimes Z^{\mathrm{st}})_i^k &= \sum_j (Z^{\mathrm{st}})_i^j \otimes (Z^{\mathrm{st}})_j^k \\
&= \sum_j (-1)^{\hat{\jmath}(\hat{\imath}+1)} z_j^i \otimes (-1)^{\hat{k}(\hat{\jmath}+1)} z_k^j.
\end{aligned}$$

Comparing signs, we get the implication $\Rightarrow$ in Eq. (1.1). The inverse implication is similar.

(b) By applying the antipode axiom 3.2(d) in Chapter 1 to Z and by using the multiplicativity, we get

$$i(Z)Z = Zi(Z) = I,$$

i.e., Eq. (1.2) for $k = 0$. Now, we have $i \circ m = m \circ S_{(12)} \circ (i \otimes i)$ (see [A] for the nongraded case). Calculating as in (a), we get from here that $i((AB)^{\mathrm{st}}) = i(B^{\mathrm{st}}) i(A^{\mathrm{st}})$ if A, B are of the same format. By applying i to Eq. $(1.2)^{\mathrm{st}}$, we get Eq. (1.2) with $k+1$ instead of k.

Finally, again as in (a), we have $S_{(12)}(Z \otimes Z)^{\mathrm{st}} = Z^{\mathrm{st}} \otimes Z^{\mathrm{st}}$. Hence, assuming that $Y = Z_k^{\mathrm{st}^k}$ is multiplicative, we can prove that $Z_{k+1}^{\mathrm{st}^{k+1}} = i(Y^{\mathrm{st}})$ is multiplicative using the identity $\Delta \circ i = S_{(12)} \circ (i \otimes i) \circ \Delta$ (cf. again [A] for the nongraded case):

$$\begin{aligned}\Delta i(Y^{\mathrm{st}}) &= S_{(12)}(i \otimes i)\Delta(Y^{\mathrm{st}}) \\ &= S_{(12)}(i(Y) \otimes i(Y))^{\mathrm{st}} = i(Y)^{\mathrm{st}} \otimes i(Y)^{\mathrm{st}}.\end{aligned}$$

1.3. LEMMA. *Let A, B^{st}, C be multiplicative matrices of the same format in a bialgebra E. Then the ideals in E, generated by the following sets, are in fact coideals:*

(a) Coefficients of $A - C$.

(b) Coefficients of $AB - I$; coefficients of $BA - I$.

Proof. (a) We have

$$\Delta(A - C) = A \otimes A - C \otimes C = A \otimes (A - C) + (A - C) \otimes C.$$

(b) In view of Lemma 1.2,

$$\begin{aligned}\Delta(AB)_i^k &= [(A \otimes A)(B^{\mathrm{st}} \otimes B^{\mathrm{st}})^{\mathrm{st}^3}]_i^k \\ &= \sum_j (A \otimes A)_i^j (B^{\mathrm{st}} \otimes B^{\mathrm{st}})_k^j (-1)^{\hat{j}(\hat{k}+1)} \\ &= \sum_{jrs} (a_i^r \otimes a_r^j)((-1)^{\hat{s}(\hat{k}+1)} b_s^k \\ &\qquad \otimes (-1)^{\hat{j}(\hat{s}+1)} b_j^s)(-1)^{\hat{j}(\hat{k}+1)} \\ &= \sum_{jrs} (a_i^r b_s^k \otimes a_r^j b_j^s) \\ &\qquad \times (-1)^{(\hat{r}+\hat{j})(\hat{s}+\hat{k})+\hat{s}(\hat{k}+1)+\hat{j}(\hat{s}+1)+\hat{j}(\hat{k}+1)} \\ &= \sum_{rs} (-1)^{\hat{r}\hat{s}+\hat{r}\hat{k}+\hat{s}\hat{k}+\hat{s}} a_i^r b_s^k \otimes (AB)_r^s.\end{aligned}$$

From this we deduce

$$\begin{aligned}\Delta(AB - I)_i^k = &\sum_{rs} (-1)^{\hat{r}\hat{s}+\hat{r}\hat{k}+\hat{s}\hat{k}+\hat{s}} a_i^r b_s^k \otimes [(AB)_s^r - \delta_r^s] \\ &+ \sum_r (a_i^r b_r^k - \delta_i^k) \otimes 1,\end{aligned}$$

which shows that the entries of $AB - I$ generate a coideal. A similar calculation can be applied to $BA - I$.

1.4. CONSTRUCTION OF THE HOPF ENVELOPE. Consider a superbialgebra E generated by coefficients of a multiplicative matrix Z. In general, E has no antipode. We interpret the problem of construction of a quantum supergroup H from a quantum supersemigroup E as a problem of construction of a universal bialgebra map $E \to H$, where H is a Hopf superalgebra. In order to construct it explicitly, put $Z = (z_i^j)$ and $E = k\langle \bar{z}_i^j \rangle / R_0$, where $\bar{z}_i^j$ are free associative variables and R_0 is the ideal of relations between z_i^j. Put $\overline{Z}_0 = (\bar{z}_i^j)$ and introduce an infinite sequence of matrices of the same format $\overline{Z}_0, \overline{Z}_1, \overline{Z}_2, \ldots$ with independent entries. Denote by $\bar{H}$ the free associative algebra generated by these entries.

Define a structure of a superbialgebra on $\bar{H}$ by

$$\bar{\Delta}(Z_k^{\mathrm{st}^k}) = \overline{Z}_k^{\mathrm{st}^k} \otimes \overline{Z}_k^{\mathrm{st}^k}, \quad \bar{\varepsilon}(\overline{Z}_k) = I; \quad k = 0, 1, 2, \ldots.$$

Define a linear map $i : \bar{H} \to \bar{H}$ by

$$i(\overline{Z}_k) = \overline{Z}_{k+1}, \qquad i(uv) = (-1)^{\hat{u}\hat{v}} i(v) i(u).$$

Denote by $\bar{R}$ the ideal in $\bar{H}$ generated by the following sets:

(1.3) the entries of $\overline{Z}_k^{\mathrm{st}^k} \overline{Z}_{k+1}^{\mathrm{st}^k} - I, \quad \overline{Z}_{k+1}^{\mathrm{st}^k} \overline{Z}_k^{\mathrm{st}^k} - I;$

(1.4) $R_k = i^k(R_0); \quad k = 0, 1, 2, \ldots.$

Put $H = \bar{H}/\bar{R}$. Clearly, there exists a superalgebra morphism $\gamma : E \to H$ defined by $\gamma(Z) = \overline{Z}_0 \bmod \bar{R}_0$.

1.4. THEOREM. *(a) $\bar{R}$ is a coideal in $\bar{H}$. Therefore, $\bar{\Delta}$ induces on H a superbialgebra structure and γ is a superbialgebra morphism.*

(b) $i(\bar{R}) \subset \bar{R}$, and i induces upon H an antipodal map so that H is a Hopf superalgebra.

(c) For an arbitrary superbialgebra morphism $\gamma' : E \to H'$, where H' is a Hopf superalgebra, there exists a unique Hopf superalgebra morphism $\beta : H \to H'$ such that $\gamma' = \beta\gamma$.

Proof. (i) *$\bar{R}$ is a coideal.* Since E is a coalgebra with $\Delta(Z) = Z \otimes Z$, we have $\bar{\Delta}(R_0) \subset k\langle \bar{z}_i^j \rangle \otimes R_0 + R_0 \otimes k\langle \bar{z}_i^j \rangle$. It follows that the same is true for R_j with even j instead of R_0. One can treat odd j similarly by using $(E, m^{\mathrm{op}}, \Delta^{\mathrm{op}})$ instead of (E, m, Δ).

In order to treat Eq. (1.3), put $A = \overline{Z}_k^{\mathrm{st}^{k+1}}$, $B = \overline{Z}_{k+1}^{\mathrm{st}^k}$. According to Lemma 1.2,

$$\bar{\Delta}(A) = A \otimes A, \quad \bar{\Delta}(B^{\mathrm{st}}) = B^{\mathrm{st}} \otimes B^{\mathrm{st}}.$$

Hence, we can apply Lemma 1.3.

(ii) $\bar{R}$ *is* i*-stable*. Clearly, $i(R_k) = R_{k+1}$. Furthermore, since i reverses multiplication, we have $i((AB)^{\mathrm{st}}) = i(B^{\mathrm{st}})i(A^{\mathrm{st}})$ so that

$$i(\overline{Z}_k^{\mathrm{st}^k}\overline{Z}_{k+1}^{\mathrm{st}^k} - I) = (\overline{Z}_{k+2}^{\mathrm{st}^{k+1}}\overline{Z}_{k+1}^{\mathrm{st}^{k+1}} - I)^{\mathrm{st}^3}.$$

(iii) $i \bmod \bar{R}$ *is an antipode*. The entries of $\overline{Z}_k^{\mathrm{st}^k}$, $k = 0, 1, 2, \ldots \bmod \bar{R}$ generate $H = \bar{H}/\bar{R}$ as a k-algebra. By reducing Eq. (1.3) $\bmod \bar{R}$ and by applying the antipode axiom to any of these generators, we see that we get an identity.

Thus, it suffices to check that if this axiom is satisfied when applied to $u, v \in H$, it is also satisfied for uv. In fact,

$$\begin{aligned} m(i \otimes \mathrm{id})(\Delta(u)\Delta(v)) &= m(i \otimes \mathrm{id})(\sum_k u_k' \otimes u_k'')(\sum_l v_l' \otimes v_l'') \\ &= m \sum_{kl} (-1)^{\hat{u}''k\hat{v}'l + \hat{u}'k\hat{v}'l} i(v_l')i(u_k') \\ &\quad \otimes u_k''v_l'' \\ &= \sum_k i(v_l')(\sum_l (-1)^{\hat{u}\hat{v}'l} i(u_k')u_k'')v_l'' \\ &= \eta\varepsilon(u)(\sum i(v_l')v_l''(-1)^{\hat{u}\hat{v}_l'}) = \eta\varepsilon(u)\eta\varepsilon(v). \end{aligned}$$

The last equality is valid even if $\hat{u} = 1$, because then $\varepsilon(u) = 0$.

(iv) *Universality.* Let $\gamma' : E \longrightarrow (H', i')$ be a superbialgebra morphism to a Hopf superalgebra. Put $Z_k' = (i')^k(\gamma'(Z))$. In view of Lemma 1.2, the entries of Z_k' verify the relations defined by Eqs. (1.3), (1.4). This allows us to define β.

1.5. REMARKS.

1.5.a. Bijective and Unipotent Antipodes. The class of Hopf superalgebras with a bijective antipode is closer to that of usual algebraic groups. It is not difficult to modify our construction in order to obtain a universal map $E \longrightarrow H$ in this class. It suffices to introduce the matrices $\overline{Z}_k$ and relations (1.3), (1.4) for all integers k.

If we impose the additional relations $\overline{Z}_k = \overline{Z}_{k+2d}$, we shall obtain a universal map into a Hopf superalgebra with the antipode verifying $i^{2d} = id$.

1.5.b. General Case. Any (super)coalgebra (E, Δ, ε) is a union of linear subspaces generated by entries of multiplicative matrices (see [A]). Hence any (super)bialgebra is a union of subbialgebras generated by entries of a multiplicative matrix. Therefore, the universal maps $E \longrightarrow H$ of the kind we have considered exist for arbitrary E and can be constructed as inductive limits.

2. Automorphisms of Quantum Spaces

2.1. QUANTUM (SUPER)SPACES. Consider the category of pairs $\mathcal{A} = (A, A_1)$, where A is a k-(super)algebra, $A_1 \subseteq A$ a finite-dimensional subsuperspace generating A. A morphism $(A, A_1) \to (B, B_1)$ is a morphism of k-superalgebras $f : A \to B$ such that $f(A_1) \subseteq B_1$. We shall consider objects of the dual category as a version of "quantum linear spaces"; the coordinate functions A_1 on such a space can be subject to arbitrary "commutation relations" represented by the (generators of the) kernel of the canonical map $T(A_1) \to A$, where $T(A_1)$ is the tensor algebra of A_1.

EXAMPLES. (a) $\dim(A) < \infty$, $A_1 = A$.

(b) A is a $\mathbb{Z}$-graded k-algebra $\oplus A_i$, with $A_0 = k$, generated by A_1.

(c) The same as in (b), with the ideal of relations generated by quadratic ones.

Let a quantum (super)group be given represented by its function Hopf algebra H. *A (left) action of this group on the quantum (super) space* (A, A_1) is the structure of a (left) H-comodule on A such that the coaction map $\delta : A \to H \otimes A$ is a morphism of superalgebras and $\delta(A_1) \subseteq H \otimes A_1$. One of the main results of this section will be the following theorem (we omit "super" for brevity).

2.2. THEOREM. *For any quantum space $\mathcal{A} = (A, A_1)$, there exists a universal action of a quantum group G on $\mathcal{A}$, represented by a coaction $\delta : A \to G \otimes A$ such that for any coaction $\delta' : A \to H \otimes A$ with $\delta'(A_1) \subseteq H \otimes A_1$, there is a morphism of Hopf algebras $\gamma : H \to G$ such that $\delta = (\gamma \otimes \mathrm{id}) \circ \delta'$.*

We shall complete the proof of a stronger result in Section 2.5. Namely, in the universality statement, H may be an arbitrary associative k-algebra and δ' an arbitrary algebra morphism (then γ is an algebra morphism). The group G acting upon $\mathcal{A}$ with such universality properties deserves the name of "quantum $\mathrm{GL}(\mathcal{A})$."

We shall start, however, with "quantum linear morphism spaces."

2.3. PROPOSITION. *For any pair of quantum spaces $\mathcal{A} = (A, A_1)$, $\mathcal{B} = (B, B_1)$, consider the category of diagrams $\varphi : A \to F \otimes B$ where F is a k-algebra and φ is a morphism of algebras satisfying $\varphi(A_1) \subseteq B_1$. A morphism of such diagrams $(F, \varphi) \to (F', \varphi')$ is an algebra morphism $F \to F'$ such that $(\alpha \otimes \mathrm{id}_B) \circ \varphi = \varphi'$.*

This category has an initial object

$$\delta = \delta_{\mathcal{A},\mathcal{B}} : A \to E \otimes B.$$

Proof. Let $A_1 = \oplus kx_i$, $B_1 = \oplus ky_j$. Any diagram (F, φ) is defined by $z_i^j \in F$ such that $\varphi(x_i) = \sum_j z_i^j \otimes y_j$. Let $\bar{x}_i$ be images of x_i in $T(A_1) = \bar{A}$, $\bar{z}_i^j$ independent associative variables, $\overline{E} = k\langle \bar{z}_i^j \rangle$; define $\bar{\varphi} : \bar{A} \to \overline{E} \otimes B$ by $\bar{\varphi}(\bar{x}_i) = \sum \bar{z}_i^j \otimes y_j$. (Parity of $\bar{z}_i^j$ is $\bar{x}_i + \bar{y}_j$.)

For a relation $f \in \mathrm{Ker}(T(A_1) \to A)$, we can write

$$\bar{\varphi}(f), = \sum g_\mu(f) \otimes Y^\mu,$$

where $\{Y^\mu\}$ is a fixed basis of B consisting, for example, of monomials in $\{y_j\}$.

Clearly, $g_\mu(f) \in k\langle \bar{z}_i^j \rangle$ vanish if one replaces $\bar{z}_i^j$ by z_i^j. One easily sees that $E = k\langle \bar{z}_i^j \rangle / (g_\mu(f))$ and that $\bar{\varphi}$ reduced modulo the same ideal furnish our initial object.

REMARK. It follows from the proof that E has a canonical linear subspace generating it: E_1 = image of $\otimes k\bar{z}_i^j$. Therefore, E is a part of a quantum space $\mathcal{E} = \mathcal{E}(\mathcal{A}, \mathcal{B}) = (E, E_1)$, the space of functions on the quantum space of morphisms "Spec" $\mathcal{B} \to$ "Spec" $\mathcal{A}$.

2.4. COMPOSITION AND IDENTITY. Of course, $\mathcal{E}(\mathcal{A}, \mathcal{B})$ have the formal properties of (co)homomorphisms objects. Looking at the composition

$$A \xrightarrow{\delta_{A,B}} E(A,B) \otimes B \xrightarrow{\mathrm{id} \otimes \delta_{B,C}} E(A,B) \otimes E(B,C) \otimes C$$

and by using the universality of $E(A, C)$, we obtain the comultiplication morphisms

$$\Delta_{\mathcal{A},\mathcal{B},\mathcal{C}} : E(A,C) \to E(A,B) \otimes E(B,C)$$

with obvious (co)associativity properties (we denote by $E(A, C)$ the algebra of $\mathcal{E}(\mathcal{A}, \mathcal{C})$, etc.).

In particular,

$$\Delta = \Delta_{\mathcal{A},\mathcal{A},\mathcal{A}} : E(A,A) \to E(A,A) \otimes E(A,A)$$

defines upon E the structure of a bialgebra. The counit map stems from the universality property applied to the diagram

$$A \xrightarrow{\mathrm{id}} k \otimes A$$

("identity constraint"). Finally, the structure morphism

$$\delta_{\mathcal{A}} : A \to E(A,A) \otimes A$$

defines on A the structure of a left comodule such that $\delta(A_1) \subseteq E \otimes A_1$.

2.5. QUANTUM GENERAL LINEAR GROUP. Denote now by $h : E = E(A, A) \to G$ the Hopf envelope of E constructed in Section 1. Clearly,

$$\delta = (h \otimes \mathrm{id}_A) \circ \delta_{\mathcal{A}} : A \to G \otimes A$$

defines an action of the quantum group "Spec" G on $\mathcal{A}$. We have seen already that any action $A \to H \otimes A$ is "induced" from δ by an algebra morphism $H \to E(A, A) \xrightarrow{h} G$. We leave to the reader an easy verification that this is, in fact, a bialgebra morphism. Finally, any bialgebra morphism of Hopf algebras commutes with antipodes because an antipode transforms any multiplicative matrix into its inverse.

2.6. EXAMPLE: QUADRATIC SPACES. Let $\mathcal{A} = (A, A_1)$ and $\mathcal{B} = (B, B_1)$ be two quadratic algebras, i.e., $\mathbb{Z}$-graded pure even algebras generated of the form $T(A_1)/(R(A))$, where $R(A)$ is a subspace in $A_1^{\otimes 2}$. In [Ma2] and [Ma4], we proved that $\mathcal{E}(\mathcal{A}, \mathcal{B})$ is also a quadratic algebra generated by $B_1^* \otimes A_1$ with the space of quadratic relations

$$S_{(23)}(R(B)^{\perp} \otimes R(A)) \subseteq (B_1^* \otimes A_1)^{\otimes 2}.$$

In fact, the category of quadratic algebras is a tensor category with inner $\underline{\mathrm{Hom}}$, unit object, and natural analogs of direct sum, symmetric powers, and exterior powers (cf. [Ma2]).

We leave to the reader a similar discussion of the category of quadratic superalgebras where new natural functors emerge, e.g., the functor Π of parity change of the generators. It will be introduced in the next section of a particular case.

2.7. EXAMPLE. $\mathrm{GL}_q(2)$. Here we shall sketch proofs of the properties of $M_q(2)$ and $\mathrm{GL}_q(2)$ stated in Chapter 1, Section 3.8. An important lesson is that in order to obtain quantum groups of reasonable size (with a function ring of polynomial growth), it is necessary to consider either multiplicative matrices coacting upon a quadratic ring A and its dual ring $A^!$ simultaneously, or to impose on this matrix an additional condition that its transpose also act upon A. This is discussed in more detail in [Ma2].

2.8. ORIGIN OF THE COMMUTATION RELATIONS $(3.5)_q$ OF CHAPTER 1. Consider the quantum space $A_q^{2|0} = (k[x, y], kx + ky)$, where $xy = q^{-1}yx$. Put $x' = a \otimes x + b \otimes y$, $y' = c \otimes x + d \otimes y$. Calculating as in the proof of Proposition 2.3, we see that the equality $x'y' = q^{-1}y'x'$ is equivalent to the commutation relations

$$(2.1) \qquad ac = q^{-1}ca; \qquad bd = q^{-1}db; \qquad ad - da = q^{-1}cb - qbc.$$

They constitute only half of the relations $(3.5)_q$ in Chapter 1 and determine the matrix quantum space $E(A_q^{2|0}, E_q^{2|0})$ in the notation of Section 2.4.

In order to obtain another half, one can proceed in two different ways.

One is to postulate that *the same matrix* $Z = \begin{pmatrix} a & b \\ c & d \end{pmatrix}$ defines a coaction $A_q^{0|2} \to k[a,b,c,d] \otimes A_q^{0|2}$, where $A_q^{0|2} = k[\xi, \eta]$, $\xi^2 = \eta^2 = \xi\eta + q\eta\xi = 0$. If $q^2 \neq -1$, these additional relations together with the previous ones are equivalent to Eq. $(3.5)_q$ in Chapter 1 and define $M_q(2)$. If $q^2 = -1$, they are slightly weaker and define a quadratic space whose homogeneous components grow like those of a 5-dimensional commutative graded ring.

This way of obtaining additional relations is quite useful, since it immediately shows the origin of DET_q and its multiplicativity: We have $A_q^{0|2} = k \oplus k\xi \oplus k\eta \oplus k\xi\eta$, and DET_q corresponds to the one-dimensional comodule $k\xi\eta$. It has, however, a drawback: An invariant definition of $A_q^{0|2}$ uses a duality functor specific for quadratic algebras, and one does not readily see what to do for a general quantum space.

Another way to obtain *the same* additional relations is to postulate Eq. (2.1) also for the transposed matrix $Z^t = \begin{pmatrix} a & c \\ b & d \end{pmatrix}$. This furnishes an important involutive isomorphism

$$\tau : (M_q(2), m, \Delta) \to (M_q(2), m, \Delta^{\mathrm{op}}); \quad \tau(Z) = Z^t,$$

(and similarly for $\mathrm{GL}_q(2)$) which allows us to identify right and left comodules (or rather to identify one category with the opposite of the other one). On the level of $\mathrm{GL}_q(2)$, one may combine τ and the antipodal maps. This shows that the reversal of either multiplicative or comultiplication, or both, in the Hopf algebra of $\mathrm{GL}_q(2)$ leads to canonically isomorphic Hopf algebras. Perhaps, a variant of this property should be postulated for "good" quantum groups.

The last question to be discussed is the choice of a basis in A_1, the generator space of (A, A_1). Nothing depends on it as long as we consider only $\mathrm{GL}(A, A_1)$. However, the ideal, generated by the relations imposed on Z and Z^t together, in general will change after a linear transformation of an initial basis. It seems that there are distinguished choices. One guess is that good choices should be defined in terms of the representation of the automorphism group of (A, A_1) (linear on A_1). This is, in fact, a group scheme over k whose function ring is that of $\mathrm{GL}(A, A_1)$ made commutative, that is, factorized by the ideal generated by commutators. For example, if $q^2 \neq 1$, $\mathrm{Aut}(A_q^{2|0}) = \mathbb{G}_m \times \mathbb{G}_m$, and the subspaces kx, ky are just the invariant subspaces of the action of this group.

2.9. COMMUTATION RELATIONS FOR Z^n. We shall now sketch a proof due to Tunstall that if Z satisfies Eq. $(3.5)_q$ in Chapter 1, then Z^n satisfies

Eq. $(3.5)_{q^n}$ in Chapter 1 for all $n \in \mathbb{Z}$. This proof is an elegant arrangement of straightforward inductive reasoning. Unfortunately, it does not explain the real source of this phenomenon, so that it is not clear how to generalize it.

For $n = -1$, this statement is true; thus, it suffices to consider only nonnegative n. Put

$$\sigma_q = \begin{pmatrix} q^{1/2} & 0 \\ 0 & q^{-1/2} \end{pmatrix}; \qquad Z^{t(q)} = \sigma_q^{-1} Z^t \sigma_q.$$

The relations $(3.5)_q$ can be written in the form

$$Z \varepsilon Z^{t(q)} = Z^{t(q)} \varepsilon Z = \mathrm{DET}_q\, \varepsilon; \quad \varepsilon = \begin{pmatrix} 0 & 1 \\ -1 & 0 \end{pmatrix},$$

suggesting that $\mathrm{SL}_q(2)$ is also $\mathrm{Sp}_q(2)$. Hence, to prove that Z^n satisfies Eq. $(3.5)_{q^n}$, it suffices to check that

$$(Z^n)^{t(q^n)} = (Z^{t(q)})^n.$$

A straightforward induction by n shows that there exist two sequences of associative polynomials $f_n(x,y,z,q)$ and $g_n(x,y,z,q)$ with the following properties:

(a) $g_0 = 1, \quad f_0 = 0;$

(b) $f_n(a,d,bc,q) = q^{1-n} f_n(d,a,bc,q^{-1});$

$(dq-a) f_n(a,d,bc,q) - q^n g_n(d,a,bc,q^{-1}) + g_n(a,d,bc,q) = 0.$

(c) $Z^n = \begin{pmatrix} g_n(a,d,bc,q) & b f_n(d,a,bc,q^{-1}) \\ c f_n(a,d,bc,q) & g_n(d,a,bc,q^{-1}) \end{pmatrix}.$

If we denote

$$g_n' = g_n(d,a,bc,q^{-1}), \qquad f_n' = f_n(d,a,bc,q^{-1}),$$

we finally find

$$(Z^n)^{t(q^n)} = (\sigma_{q^n})^{-1} (Z^n)^t \sigma_{q^n} = \begin{pmatrix} g_n & q^{-1} c f_n' \\ q b f_n & g_n' \end{pmatrix};$$

$$(Z^{t(q)})^n = \begin{pmatrix} a & q^{-1} & c \\ qb & d & \end{pmatrix}^n = \begin{pmatrix} g_n & q^{-1} c f_n' \\ q b f_n & g_n' \end{pmatrix}.$$

2.10. THE CARTAN SUBGROUP OF $\mathrm{SL}_q(2)$. A one-dimensional algebraic torus is represented by the commutative Hopf algebra $T = k[t,t^{-1}]$,

$\Delta(t) = t\otimes t$, $\varepsilon(t) = 1$, $i(t) = t^{-1}$. Hence, it is natural to define a Cartan subgroup as a surjective morphism $G \to T$, where G is the function algebra of $\mathrm{SL}_q(2)$. But such a morphism can be factored through $c : G \to G/[G,G] = k[\bar{a},\bar{d}]/(\bar{a}\bar{d} - 1)$, where $\bar{a} = a \bmod [G,G]$, $\bar{d} = d \bmod [G,G]$. This is itself a Cartan subgroup. Thus, after the quantization, precisely one Cartan subgroup survives (if $q^2 \neq 1$), and the reason is that the quantum plane $A_q^{2|0}$ still has $\mathbb{G}_m \times \mathbb{G}_m$ as its automorphism group.

T can be used to develop an elementary root technique. For example, the regular corepresentation $\Delta : G \to G\otimes G$ can be decomposed with respect to the right and left coaction of T in the following way. Put $R = (\mathrm{id}_G \otimes c)\circ\Delta : G \to G\otimes T$; $L = (c\otimes \mathrm{id}_G)\circ\Delta : G \to T\otimes G$ and further

$$G[m,n] = \{f \in G | L(f) = t^m \otimes f, R(f) = f\otimes t^n\}.$$

Then one can prove that $G[0,0] = k[q^{-1}bc]$ and $G[m,n]$ are free left and right modules of rank 1 over $G[0,0]$ whose generators can be calculated explicitly as monomials of a, b, c, d.

2.11. SOME PROPERTIES OF COMODULES. The standard simple comodules of $\mathrm{SL}(2)$ have their natural quantum analogs: They are simply homogeneous components of $A_q^{2|0}$, that is, quantum symmetric powers of the fundamental representation. They can be embedded into the regular corepresentation as

$$V_s^L = k[a,c]_{(2s)}; \qquad V_s^R = k[a,b]_{(2s)},$$

where s is the spin, $s \in \frac{1}{2}\mathbb{Z}_+$; V^L (resp. V^R) is a left (resp. a right) subcomodule of G. Let Z_s be the multiplicative matrix corresponding to a choice of basis in V_s^L. Denote by W_s the linear span of its elements in G. This is a bicomodule.

If q is not a root of unity, and G is the direct sum of W_s, there is a natural isomorphism of bicomodules $V_s^L \otimes V_s^R \to W_s$. Furthermore, any simple right (resp. left) comodule is isomorphic to V_s^R (resp. V_s^L) for some s.

The results of Sections 2.10 and 2.11, as well as of Section 2.12 below, are proved in [MMNNU]; see also [VS] for some closely related information.

2.12. INTEGRAL AND COMPLETE REDUCIBILITY. Let H be a Hopf algebra. A linear functional $f : H \to k$ is called a left- (resp. right-) invariant integral iff $(f\otimes \mathrm{id})\circ\Delta = \eta\circ f$ (resp. $(\mathrm{id}\otimes f)\circ\Delta = \eta\circ f$). The following result, proved in [A], shows the role of this notion.

2.13. THEOREM. *(a) The following properties of H are equivalent:*

(i) Any left (resp. right) H-comodule is completely reducible.

(ii) H as a left (resp. right) H-comodule is completely reducible.

(iii) H also admits a left- (resp. right-) invariant integral.

(iv) There exists an invariant scalar product $s : H \otimes H \to k$, that is, a bilinear form on H with the following properties:

$$(\mathrm{id} \otimes s) \circ (\Delta \otimes \mathrm{id}) = (s \otimes \mathrm{id}) \circ (\mathrm{id} \otimes \Delta); \quad s \circ \Delta = \varepsilon.$$

(b) There is no more than one integral (up to a scalar).

Sketch of proof of (a). (i) $\Rightarrow$ (ii). Clear.

(ii) $\Rightarrow$ (iii). An integral is a splitting of the comodule morphism $\eta : k \to H$, where the structure of a comodule on k is again given by $\eta : k \to k \otimes H$ (resp. $k \to H \otimes k$).

(iii) $\Rightarrow$ (iv). Let $f : H \to k$ be a left integral, $f(1) = 1$. Then s can be defined explicitly by $s = f \circ m \circ (\mathrm{id} \otimes \Delta)$.

(iv) $\Rightarrow$ (i). Assume s is given. Let $j : N \to M$ be an injective morphism of left H-comodules. We want to construct a projector $q : M \to N$. We start with a linear projector $p : M \to N$ and put

$$q = (s \otimes \mathrm{id}_N) \circ (\mathrm{id}_H \otimes \delta_N) \circ (\mathrm{id}_H \otimes p) \circ \delta_M.$$

Turning now to the case $\mathrm{SL}_q(2)$, one can try to calculate an invariant integral explicitly. In [MMNNU] the following is proved: if q is not a root of unity, f exists if $f(1) = 1$, we have:

(a) $f|G[m,n] = 0$ for $(m,n) \neq (0,0)$.

(b) $f((bc)^n) = (-q)^{-n}(1 - q^2)(1 - q^{2(n+1)})^{-1}$.

(c) $f \circ i = f$.

From the proof one can deduce that f does not exist when q is a root of unity of a sufficiently high degree. I do not know what happens if $q^2 = 1$; recall that for $q^2 = -1$, the group $\mathrm{GL}_q(2)$ can be naturally embedded into a larger quantum group (less commutation relations). In [Lu2] and [Lu3], the case $q^n = 1$ is treated for all Drinfeld–Jimbo groups, and very interesting parallels with the classical theory in the char. p case are suggested.

3. General Linear Supergroups

3.1. DEFORMATION PARAMETERS AND COMMUTATION RELATIONS. In this section, based on [Ma5], we apply the method of Section 2 to the construction and investigation of the simplest analogs of general linear supergroups that represent the quantum automorphisms of the commutation relations of the type $x_i x_j = q_{ij} x_j x_i$. We discover that the Poincaré–Birkhoff–Witt-type theorem is valid not only for the traditional deformation $\mathrm{GL}_q(n)$

(and its superanalog) corresponding to $q_{ij} = q$, but also for $q_{ij} = \varepsilon_{ij} q$, $\varepsilon_{ij} = \pm 1$. We also construct the quantum Berezinian for these supergroups.

Let us start by stating the commutation relations. From now on, we fix a format $\{a_1, \ldots, a_n\}$ and a family $q = \{q_{ij} | 1 \leq i, j \leq n\}$ of nonzero elements of k. We put $\hat{\imath} = a_i$ and denote by E_q the algebra generated by n^2 symbols z_i^k subject to the following relations:

$$(3.1) \quad (z_i^k)^2 = 0 \text{ for } \hat{\imath} + \hat{k} \text{ odd };$$

$$(3.2) \quad z_i^k z_i^l - (-1)^{(\hat{k}+1)(\hat{l}+1)} q_{kl} z_i^l z_i^k = 0 \text{ for } \hat{\imath} \text{ odd}, \quad k < l;$$

$$(3.3) \quad z_i^k z_i^l - (-1)^{\hat{k}\hat{l}} q_{kl}^{-1} z_i^l z_i^k = 0 \text{ for } \hat{\imath} \text{ even}, \quad k < l;$$

$$(3.4) \quad z_i^k z_j^k - (-1)^{\hat{\imath}\hat{\jmath}} q_{ij}^{-1} z_j^k z_i^k = 0 \text{ for } \hat{k} \text{ even}, \quad i < j;$$

$$(3.5) \quad z_i^k z_j^k - (-1)^{(\hat{\imath}+1)(\hat{\jmath}+1)} q_{ij} z_j^k z_i^k = 0 \text{ for } \hat{k} \text{ odd}, \quad i < j;$$

$$(3.6) \quad (-1)^{\hat{k}(\hat{\jmath}+\hat{l})} z_i^k z_j^l + (-1)^{\hat{\jmath}\hat{l}} q_{kl} z_i^l z_j^k = q_{ij}^{-1} (-1)^{\hat{\imath}\hat{\jmath}} [(-1)^{\hat{k}(\hat{\imath}+\hat{l})} z_j^k z_i^l + (-1)^{\hat{\imath}\hat{l}} q_{kl} z_j^l z_i^k];$$

$$(3.7) \quad (-1)^{(\hat{k}+1)(\hat{\jmath}+\hat{l})} z_i^k z_j^l + (-1)^{(\hat{\jmath}+1)(\hat{l}+1)} q_{kl}^{-1} z_i^l z_j^k = q_{ij} (-1)^{(\hat{\imath}+1)(\hat{\jmath}+1)} [(-1)^{(\hat{k}+1)(\hat{\imath}+\hat{l})} z_j^k z_i^l + (-1)^{(\hat{\imath}+1)(\hat{l}+1)} q_{kl}^{-1} z_j^l z_i^k] \quad \text{for} \quad i < j, \quad k < l.$$

We define a $\mathbb{Z}$-grading of E_q by $z_i^k = \hat{\imath} + \hat{k}$ so that $Z = (z_i^k)$ has the format $(a_1, \ldots, a_n)$.

Now we can state our first main result.

3.2. THEOREM. *There is a unique structure of a superalgebra on E_q for which Z is multiplicative.*

Notice, that if one puts $q_{ij} = 1$ for all i, j, then Eqs. (3.1)–(3.7) become equivalent to the simple supercommutation relations $[z_i^j, z_k^l] = 0$. Therefore, E_q is a deformation of the ring of polynomial functions on the supermanifold $\mathrm{Mat}(a|b)$, where a (resp. b) is the number of even (resp. odd) a_i in our format.

We shall show that E_q is a universal bialgebra co-acting upon two dual "quantum superspaces" A_q and A_q^*.

QUANTUM SUPERSPACE A_q. This space (or, rather, polynomial function ring on it) is generated by the coordinates $x_1, \ldots, x_n$ with parity assignment $x_i = \hat{\imath}$ and commutation rules

$$(3.8) \quad x_i^2 = 0 \quad \text{for} \quad \hat{\imath} = 1;$$

(3.9) $\quad x_i x_j - q_{ij}^{(-1)^{\hat{i}\hat{j}}} x_j x_i = 0 \quad \text{for} \quad i < j.$

QUANTUM SUPERSPACE A_q^*. It is generated by the coordinates $\xi^1, \ldots, \xi^n$ with $\hat{\xi}^k = 1 - \hat{k}$ and commutation rules

(3.10) $\quad (\xi^k)^2 = 0 \quad \text{for} \quad \hat{k} = 0;$

(3.11) $\quad \xi^k \xi^l - q_{kl}(-1)^{(\hat{k}+1)(\hat{l}+1)} \xi^l \xi^k = 0 \quad for \quad k < l.$

Note that if we put $V = \oplus_i k x_i$, $V^* = \oplus k \xi^j$ and define an *odd* bilinear pairing $\langle\,|\,\rangle : V^* \otimes \to k$ by $\langle \xi^j | x_i \rangle = \delta_i^j$, then the left-hand side tensors in Eqs. (3.10) and (3.11) generate a subspace in $V^* \otimes V^*$ that is the orthogonal complement of the subspace in $V \otimes V$ generated by the left-hand sides of Eqs. (3.8) and (3.9) (to check it, use $\langle \xi^k \otimes \xi^l | x_i \otimes x_j \rangle = (-1)^{\hat{i}(\hat{l}+1)} \delta_i^k \delta_j^l$).

Therefore, A_q^* is different from $A_q^!$ as defined in [Ma2] and [Ma4], where an *even* bilinear pairing is used. One can say that $*$ combines ! and parity change.

Theorem 3.2 is a direct consequence of the following result.

3.3. THEOREM. *(a) There exist algebra morphisms*

$$\delta : A_q \to E_q \otimes A_q, \quad \delta^* : A_q^* \to E_q \otimes A_q^*$$

such that

(3.12) $\quad \delta(x) = Z \otimes x, \quad i.e., \quad \delta(x_i) = \sum_{j=1}^{n} z_i^j \otimes x_j;$

(3.13) $\quad \delta^*(\xi) = Z \otimes \xi, \quad i.e., \quad \delta^*(\xi^k) = \sum_{k=1}^{n} z_k^l \otimes \xi^l.$

(b) The pair (δ, δ^*) is universal in the following sense. Let B be a superalgebra and let (γ, γ^*) be two superalgebra morphisms

$$\gamma : A_q \to B \otimes A_q, \qquad \gamma^* : A_q^* \to B \otimes A_q^*$$

such that $\gamma(V) \subseteq B \otimes V$, $\gamma^*(V^*) \subseteq B \otimes V^*$, and

$$\gamma(x) = Y \otimes x, \qquad \gamma^*(\xi) = Y \otimes \xi$$

for some $Y \in \mathrm{Mat}(n, B)$ of the same format $(\hat{\imath})$ as Z. Then there exists a unique superalgebra morphism $\beta : E_q \to B$ such that

$$\gamma = (\beta \otimes \mathrm{id}) \circ \delta, \qquad \gamma^* = (\beta \otimes \mathrm{id}) \circ \delta^*.$$

Proof. As in the proof of Proposition 2.3, we start with an arbitrary superalgebra morphism $\gamma : A_q \to B \otimes A_q$, linear in x_i, and temporarily denote by $z_i^k \in B$ the coefficients in

$$\gamma(x_i) = \sum_{j=1}^{n} z_i^j \otimes x_j.$$

Clearly, γ is well-defined iff it conserves the identities (3.8) and (3.9). We must apply γ to the left-hand side of these identities and then calculate the coefficients of the resulting expressions represented as linear combinations of the monomials $1 \otimes x_i x_j$, $i < j$, which are linearly independent over B in $B \otimes A_q$. In this way, we shall obtain one half of the commutation relations (3.1)-(3.7), namely:

coefficient of x_k^2, $\hat{k} = 0$, in $\gamma(3.8) \Rightarrow (3.1)$ for $\hat{\imath} = 1$, $\hat{k} = 0$;
coefficient of $x_k x_l$, $k < l$, $\hat{\imath} = 1$ in $\gamma(3.8) \Rightarrow (3.2)$;
coefficient of x_k^2, $\hat{k} = 0$, in $\gamma(3.9) \Rightarrow (3.4)$;
coefficient of $x_k x_l$, $k < l$, in $\gamma(3.9) \Rightarrow (3.6)$.

The other half of our relations is supplied by $\gamma^* : A_q^* \to B \otimes A_q^*$. It can be directly deduced from the previously established relations by applying to them the parity change $\hat{\imath} \to 1 - \hat{\imath}$ and the parameter change $q_{ij} \to q_{ij}^{-1}$.

This calculation simultaneously proves both parts of Section 3.3 and Theorem 3.2.

By definition, $\mathrm{GL}_{(q)}(f)$ is the quantum supergroup whose function ring is the Hopf envelope H_q of E_q (here f is the format of Z).

3.4. Fundamental Corepresentations and Their Symmetric Powers. For each $m \geq 0$, the space of forms of degree m in A_q plays the role of the mth symmetric power of the fundamental corepresentation $V = \oplus k x_i$. Similarly, we have the "symmetric quantum power" of the odd-contragradient corepresentation $V^* = \oplus k \xi^j$.

3.5. Quantum Determinant. As was explained in [Ma2] and [Ma4], there is a natural construction of the quantum determinant of the matrix Z generating E_q when it is of the formal $(1, \ldots, 1)$, i.e., when A_q is a deformed exterior algebra of n indeterminates $x_1, \ldots, x_n$. We simply define $D = \mathrm{DET}_q(Z) \in E_q$ by the identity

$$\Delta(x_1, \ldots, x_n) = D \otimes x_1, \ldots, x_n \in E_q \otimes A_q,$$

which exists since $x_1, \ldots, x_n$ generates the degree-n component of A.

From this definition, one sees that $\Delta(D) = D \otimes D$, and

$$\mathrm{DET}_q(Z) = \sum \mathrm{sgn}(i_1, \dots, i_n) \prod_{\substack{a>b \\ i_a<i_b}} q_{i_a i_b} z_1^{i_1} \dots z_n^{i_n}.$$

This definition is based upon the existence of a natural one-dimensional comodule for E_q, the quantum highest exterior power $(A_q)_n$ of the fundamental corepresentation.

In the general case, it is necessary to construct the quantum Berezinian. Since it is not polynomial in entries of Z even in the classical case, one cannot expect it to lie in E_q. It may lie in H_q where Z becomes invertible. However, in order to construct the one-dimensional corepresentation furnishing our Berezinian, I shall have to diminish H_q further by making an additional factorization. I do not know whether it is really necessary.

3.6. Quantum Koszul Complex. Put $B_q = A_q^* \otimes A_q$. Clearly, this is an algebra of the same kind as A_q, with generators $(\xi^1 \otimes 1, \dots, \xi^n \otimes 1, 1 \otimes x_1, \dots, 1 \otimes x_n)$ of format $(1 - a_1, \dots, 1 - a_n, a_1, \dots, a_n)$. Put

$$c = \sum_{i=1}^{n} \xi^i \otimes x_i \in B_q.$$

One immediately verifies that $c^2 = 0$. Let $d : B_q \to B_q$ be the linear map $df = fc$. Put $H^{\cdot}(B_q) = \mathrm{Ker}(d)/\mathrm{Im}(d)$.

3.7. Proposition. *$H^{\cdot}(B_q)$ is a one-dimensional k-space generated by*

$$\prod_{\hat{\imath}=0} \xi^i \prod_{\hat{\jmath}=1} x_j \bmod \mathrm{Im}(d).$$

This space will be our superanalog of the highest exterior power. We must now prepare a Hopf superalgebra coacting upon it.

Let us start with the universal Hopf algebra G_q coacting upon B_q, $\delta : B_q \to G_q \otimes B_q$. Put

$$\delta\begin{pmatrix} \xi \\ x \end{pmatrix} = \begin{pmatrix} U & W \\ W' & V \end{pmatrix} \otimes \begin{pmatrix} \xi \\ x \end{pmatrix}; \qquad \xi = \begin{pmatrix} \xi^1 \\ \vdots \\ \xi^n \end{pmatrix}, \quad x = \begin{pmatrix} x_1 \\ \vdots \\ x_n \end{pmatrix}.$$

3.8. Theorem. *(a) The entries of the matrices $i^n(W)$, $i^n(W')$, and $i^n(V - i(U^{\mathrm{st}}))$, $n = 0, 1, 2, \dots$ generate an i-stable coideal S in G_q. Put $F_q = G_q/S$ with natural coaction upon B_q, which we denote by δ_F.*

(b) $\delta_F(c) = 1 \otimes c \in F_q \otimes B_q$. *Therefore,* δ_F *induces a coaction*

$$\delta_H : H^{\cdot}(B_q) \to H^{\cdot}(F_q \otimes B_q) = F_q \otimes H^{\cdot}(B_p).$$

(c) Put

$$\delta_H(\prod_{i=0} \xi^i \prod_{j=1} x_j) = D \otimes \prod_{i=0} \xi^i \prod_{j=1} x_j \operatorname{mod} \operatorname{Im}(d).$$

Then $\delta_H(D) = D \otimes D$; D *is called the quantum Berezinian (of* V*).*

Note that in view of the universality, there exists a Hopf superalgebra morphism $H_q \to F_q$ for which Z goes into V. If it is an embedding, we may transport D into H.

Proof of Proposition 3.7. Since the x_i, ξ^j verify Eqs. (3.8)–(3.11), one easily sees that

$$(\xi^i \otimes x_i)^2 = 0; \qquad (\xi^i \otimes x_i)(\xi^j \otimes x_j) + (\xi^j \otimes x_j)(\xi^i \otimes x_i) = 0;$$

hence, $c^2 = 0$ and even $c_\lambda^2 = 0$ for $c_\lambda = \sum \lambda_i \xi^i \otimes x_i$, $\lambda_i \in k$. For $n = 1$, B_q is a polynomial ring in one even and one odd variable so that Proposition 3.7 is evident. The general case can be reduced to this one because B_q as a complex is isomorphic to the tensor product of n classical $n = 1$ Koszul complexes. We leave the case of twisting by q_{ij} to the reader.

3.10. Proof of Theorem 3.8. First of all, S is a coideal. In fact,

$$\Delta\begin{pmatrix} U & W \\ W' & V \end{pmatrix} = \begin{pmatrix} U & W \\ W' & V \end{pmatrix} \otimes \begin{pmatrix} U & W \\ W' & V \end{pmatrix}.$$

Hence,

$$\Delta(W) = W \otimes V + U \otimes W; \qquad \Delta(W') = W' \otimes U + V \otimes W'$$

so that the entries of W and W' generate a coideal. By applying this first to the matrix $i^n \operatorname{st}^n$, we get the same conclusion for $i^n(W)$, $i^n(W')$.

If A, B are two multiplicative matrices of the same format, the coefficients of A–B generate a coideal (Lemma 3.3). By applying this to $A = i^n(V^{\operatorname{st}^n})$, $B = i^{n+1}(V^{\operatorname{st}^{n+1}})$, we see that S is a coideal.

It remains to check that c is a coinvariant. Put $\xi^t = (\xi^1, \dots, \xi^n)$. We have

$$\begin{aligned} \delta_F(\xi^t) &= (U \otimes \xi + W \otimes x)^t \operatorname{mod} S \\ &= \xi^t \otimes U^{\operatorname{st}} \operatorname{mod} S = \xi^t \otimes i(V) \operatorname{mod} S. \end{aligned}$$

Therefore,

$$\delta(c) = \delta(\xi^t x) = \xi^t \otimes i(V)V \bmod S \otimes x = 1 \otimes c.$$

The rest is clear.

3.11. THE STRUCTURE OF E_q. It is pretty clear that A_q (resp. A_q^*) is "of the same size" as a polynomial ring of a even and b odd (resp. a odd and b even) supercommuting variables. This means that the normally ordered monomials, say, $x_1^{m_1} \dots x_n^{m_n}$, with $m_i = 0, 1, 2, \dots$ for $\hat{\imath} = 0$ and $m_i = 0, 1$ for $\hat{\imath} = 1$, form a k-basis of A_q.

It is known that when the analogous statement is true for E_q when all q_{ij} are equal and when there are no odd variables (in the dual language, this is equivalent to the quantum Poincaré-Birkhoff-Witt theorem). On the other hand, it is false for general values of the parameters. We prove the following result, which gives a fairly complete picture.

Define an ordering of z_i^j by

$$z_i^j < z_k^l \quad \text{if either} \quad i > k, \quad \text{or} \quad i = k,\ j > l.$$

Call a monomial in z_i^j normally ordered if for any $z' < z''$ in this monomial z' is to the left of z'', and if no odd z' enters this monomial twice.

3.12. THEOREM. *(a) Quadratic normally ordered monomials form a basis of the quadratic part of E_q iff $q_{ij}^2 \neq -1$ for all i, j.*

(b) If this condition is satisfied, then the normally ordered monomials span E_q. They are linearly independent iff cubic monomials are independent.

(c) Cubic monomials are independent iff $q_{ij} = \varepsilon_{ij} q^{\eta_{ij}}$, where $\varepsilon_{ij}, \eta_{ij} = \pm 1$, q is arbitrary and $(\eta_{ij} - \eta_{ik})(\eta_{jk} - \eta_{ik}) = 0$ for all $i < j < k$. (For $n = 2$, cubic monomials are always independent.) If this condition is satisfied, one can renumber $x_1, \dots, x_n$ in such a way that all η_{ij} become equal.

Note that a specialization of the values of q_{ij} may enlarge E_q. Somewhat paradoxically, the largest E_q for $n = 2$ corresponds to $q^2 = -1$. Therefore, from the Hopf viewpoint, the algebra $k\langle x, y\rangle / (xy - iyx)$ is more symmetric than $k[x, y]$.

3.13. Proof. The relations (1.1)–(1.7) for some fixed indices i, j, k, l constitute a block corresponding to a choice of a 2×2-submatrix in Z. In order to present them in a more manageable form, we first put

$$\begin{pmatrix} z_i^k & z_i^l \\ z_j^k & z_j^l \end{pmatrix} = \begin{pmatrix} a & b \\ c & d \end{pmatrix}; \qquad \begin{array}{ll} q_{ij} = q_v & \text{(vertical);} \\ q_{kl} = q_h & \text{(horizontal).} \end{array}$$

Then Eqs. (3.2)–(3.7) become

$$(3.14)\quad ab = (-1)^{\hat a\hat b} q_h^{(-1)^{\hat\imath+1}} ba; \qquad cd = (-1)^{\hat c\hat d} q_h^{(-1)^{\hat\jmath+1}} dc;$$

$$(3.15)\quad ac = (-1)^{\hat a\hat c} q_v^{(-1)^{\hat k+1}} ca; \qquad bd = (-1)^{\hat b\hat d} q_v^{(-1)^{\hat l+1}} db;$$

$$(3.16)\quad ad - (-1)^{\hat a\hat d} q_v^{-1} q_h da = -\eta(q_h bc - (-1)^{\hat b\hat c} q_v^{-1} cb);$$

$$(3.17)\quad ad - (-1)^{\hat a\hat d} q_v q_h^{-1} da = \eta(q_h^{-1} bc - (-1)^{\hat b\hat c} q_v cb),$$

where $\eta = (-1)^{\hat k\hat\jmath+\hat k\hat l+\hat\jmath\hat l}$. In particular, if $q_v = q_h = 1$, then a, b, c, d supercommute pairwise.

QUADRATIC MONOMIALS. Call a monomial $d^\delta c^\gamma b^\beta a^\alpha$ normally ordered if $\alpha, \beta, \gamma, \delta = 0, 1, 2, \dots$ (resp. 0, 1) for even (resp. odd) entries. Four non-normally ordered monomials ab, cd, ac, cd can be expressed via normally ordered ones with the help of Eqs. (3.14) and (3.15). The rest of them, ad and bc, can be expressed via Eqs. (3.16) and (3.17) iff

$$\det\begin{pmatrix} 1 & \eta q_h \\ 1 & -\eta q_h^{-1}\end{pmatrix} = -\eta(q_h^{-1} + q_h) \neq 0,$$

i.e., iff $q_h^2 \neq -1$. Otherwise, by subtracting Eq. (3.16) from Eq. (3.17), we get a linear relation between the normally ordered monomials

$$(-1)^{\hat a\hat d}(q_v + q_v^{-1}) q_h da = -(-1)^{\hat b\hat c}\eta(q_v + q_v^{-1}) cb,$$

which disappears if also $q_v^2 = -1$. But then Eqs. (3.16) and (3.17) become equivalent so that the normally ordered quadratic monomials cease to span the quadratic part of E_q. In fact, one should complement them by either ad or bc.

In general, if $q_h^2 \neq -1$, Eqs. (3.16) and (3.17) together are equivalent to

$$(3.18)\quad ad = \frac{q_v^{-1} + q_v}{q_h^{-1} + q_h}(-1)^{\hat a\hat d} da + \frac{q_h^{-1} q_v^{-1} - q_h q_v}{q_h^{-1} + q_h}\eta(-1)^{\hat b\hat c} cb,$$

$$(3.19)\quad bc = \frac{q_v^{-1} + q_v}{q_h^{-1} + q_h}(-1)^{\hat c\hat b} cb + \frac{q_v^{-1} q_h - q_v q_h^{-1}}{q_h^{-1} + q_h}\eta(-1)^{\hat a\hat d} da.$$

Note now that the normal ordering of the entries of Z induces the normal ordering of the entries of an arbitrary 2×2-submatrix of Z. Therefore, if $q_{ij}^2 \neq -1$ for all q_{ij}, the normally ordered quadratic monomials form a basis of the quadratic part of E_q. This proves the first assertion of Theorem 3.12.

NORMALLY ORDERED MONOMIALS SPAN E_q. Now we shall describe an algorithm I_m, $m \leq n$, allowing us to reduce any polynomial of z_i^j, $i \geq m$, to a normal form, i.e., to present it as a linear combination of the normally ordered monomials. We shall do this inductively, starting with I_n and then consecutively defining I_{n-1}, $I_{n-2,\dots,i_1}$. The last algorithm I_1 will do the job.

Each algorithm consists of applying a series of "elementary transformations." An elementary transformation substitutes a left-hand side of one of the relations (3.14)–(3.17) occurring in a polynomial by the corresponding right-hand side.

I_n acts upon polynomials by depending only on the entries of the last line of Z. Since the corresponding commutation relations have the simple form (3.14), one can simply rearrange the entries in the normal order by supplementing the twisting coefficients and by deleting all monomials where a square of an odd variable occurs.

If I_m is already defined, I_{m-1} acts in two stages. At the first stage, we take an arbitrary neighboring pair in a monomial, of the type yx, where y lies in the $(m-1)$-th line while x lies below this line. By using Eqs. (3.18), (3.19), or (3.15), we replace this pair by a linear combination of two pairs of the type $x'y'$, where y' lies in the $(m-1)$-th line while x' lies below. After a finite number of such steps, we shall arrive at a linear combination of monomials in each of which the elements of the $(m-1)$-th line lie to the right of the elements of the lines $m, m+1, \dots, n$. At the second stage, we rearrange the elements of the $(m-1)$-th line and the rest separately, using Eq. (3.14) and I_{m-1}, respectively.

This algorithm, for the case of a pure even format and for all q_{ij} being equal, was suggested by Kobyzev.

CUBIC MONOMIALS. In general, elementary transformations can be applied to a polynomial in various orders. If two different normalization procedures lead to different results, we get a linear relation between normally ordered monomials in E_q.

This may first happen in degree three. Namely, a monomial uvw can be normalized in two different ways iff we have $u > v > w$, so that we can start rearranging either uv or vw. The following lemma deals with the case when u, v, w simultaneously lie in a 2×2-submatrix of Z as at the beginning of the proof.

3.14. LEMMA. *Suppose that* $q_h^2 \neq -1$, $q_v^2 \neq -1$. *Then the normally ordered cubic monomials in* a, b, c, d *are linearly independent in* E_q *iff* $q_v = \pm q_h^{\pm 1}$.

Proof. We must directly normalize the eight monomials a^2d, ad^2, b^2d, bd^2, abc, abd, acd, and bcd in two different ways and compare the results.

We shall do that for abc and leave the rest to the reader.

Denote by R_{ij} the application of an elementary transformation to the ij-places of a monomial, e.g., $R_{23}(abc) = a$ (r.h.s of Eq. (3.19)).

By applying $R_{23}R_{12}R_{23}$ to abc, we get a polynomial in which cba enters with the coefficient

$$(-1)^{\hat{a}\hat{d}+\hat{b}\hat{c}}(q_v^{-1}q_h - q_v q_h^{-1})(q_v^{-1}q_h^{-1} - q_v q_h)(q_h^{-1} + q_h)^{-2}$$

$$+(-1)^{\hat{b}\hat{c}+\hat{a}\hat{c}+\hat{a}\hat{b}} q_v^{(-1)^{\hat{k}+1}} q_h^{(-1)^{\hat{\imath}+1}} (q_v^{-1} + q_v)(q_h^{-1} + q_h)^{-1}. \tag{3.20}$$

On the other hand, cba appears in $R_{12}R_{23}R_{12}(abc)$ with the coefficient that coincides with the second term of Eq. (3.20). Therefore, the two normal forms can coincide only if either $q_v^{-1}q_h = q_v q_h^{-1}$ or $q_v^{-1}q_h^{-1} = q_v q_h$, i.e., $q_v = \pm q_h^{\pm 1}$. Then one checks that, for this condition, the coefficients of da^2 in these two normal forms automatically coincide for $\hat{a} = 0$; otherwise, $da^2 = 0$.

UNIQUENESS OF THE NORMAL FORM. Assume now that normal cubic monomials are linearly independent in E_q. We want to deduce that monomials of any degree are linearly independent. This is equivalent to the statement that they have a unique normal form.

There is a well-known combinatorial principle allowing us to ascertain such uniqueness (cf. [Be], [ArS]).

Namely, let p be a polynomial and let T', T'' be two elementary transformations applicable to p. Suppose that there exist two sequences of elementary transformations $S_1', \dots, S_a'$ and $S_1', \dots, S_b'$ such that

$$S_a' \dots S_1' T''(p) = S_b'' \dots S_1'' T''(p).$$

If this condition is valid for all triples (p, T', T''), then the normal form is unique.

Let us check this condition in our case. In fact, if T' and T'' are either applied to different monomials in p or to the same monomial, but the respective pairs do not intersect, we can simply put $S_1' = T''$, $S_1'' = T'$. Otherwise, up to renaming, T' replaces uv, and T'' replaces vw in a neighboring triple uvw, $u > v > w$. By assumption, uvw has a unique normal form. By reducing $T'(uv)$ and $T''(vw)$ to this common normal form, we obtain S_i' and S_j''.

THE $n = 2$ CASE. Here $q_v = q_h$ so that Lemma 3.14 can be directly applied. No further cubic monomials need be checked.

THE GENERAL CASE. Here all relevant entries lie in a 3×3-submatrix of Z, and there are in general three horizontal and three vertical coefficients q_{ij}.

One must consider 22 combinatorially different relative positions of u, v, w in such a submatrix and compare the two normal forms of all monomials uvw. The calculations are the same as in the proof of Lemma 3.14 but longer. They were made by Demidov in his diploma work (Moscow University, 1989) and the result is stated in part (c)of the Theorem 3.12.

RENUMBERING OF COORDINATES. Assume now that $(\eta_{ij}-\eta_{ik})(\eta_{jk}-\eta_{ik}) = 0$ for all $i < j < k$ and not all η_{ij}, $i < j$, are equal. If we take (i,k) with $\eta_{ik} = -1$ such that $k - i$ takes the minimal possible value, this value must be 1. Therefore, by transposing x_i with x_{i+1}, we diminish the quantity of (-1)'s among η_{ij} by one. Proceeding in this way, we can make all η_{ij} equal. (This argument is due to Gelfand.)

4. Regular Quantum Spaces

4.1. DEFINITION. *A graded k-algebra* $A = \oplus_{i\geq 0} A_i$ *with* $A_0 = k$, *generated by* A_1, $\dim(A_1) < \infty$, *is called regular if:*

(i) It is of polynomial growth: $\dim(A_n) \leq cn^{\delta}$ *for some* c, δ.

(ii) It is Gorenstein, i.e., there is a finite free-graded resolution of the right A-module k_A *such that its A-dual complex is a finite free-graded resolution of the left A-module* ${}_Ak$. *The length of this resolution is called the dimension of A.*

This definition is given in [ArS] and [ArTvdB]. These papers are devoted to the classification of regular algebras of dimension ≤ 3. Actually, one- and two-dimensional algebras can be described without any difficulty, but three-dimensional ones include several algebraic families connected with elliptic curves and their automorphisms, and their understanding requires some new and interesting techniques.

Elliptic algebras emerged in [ArS] and [ArTvdB] in connection with a classification problem. They have already appeared earlier in Sklyanin's work [Sk1,2] in the context of Yang–Baxter equations and quantum groups. They were considered further in [OF1,2], where the technique of "point-modules" was also introduced, as in [ArTvdB].

We believe that quantum automorphism groups of regular quantum spaces deserve a systematic study.

Let us start with dimension two.

4.2. PROPOSITION. *Any two-dimensional regular algebra is isomorphic to* $k\langle x,y\rangle/(f)$, *where either* $f = yx - qxy$, $q \neq 0$, *or* $f = xy - yx - y^2$.

(Cf. [ArS], where this is stated without proof, but the method of classification of three-dimensional algebras leads directly to this simpler result.)

As we have seen in Section 2.8, GL_q is just the automorphism quantum group of the pair of quadratic algebras consisting of $A_q^{2|0} = k\langle x, y\rangle/(yx - qxy)$ and the dual algebra. Exceptional values of q, roots of unity, can be read off the structure of $A_q^{2|0}$: Precisely for these values, it becomes finite over its center.

Following the same plan, we shall now study quantum endomorphisms and automorphisms of the second regular algebra (or, rather, pairs consisting of this algebra and its dual). We shall call them $M_{\mathrm{J}}(2)$ and $\mathrm{GL}_{\mathrm{J}}(2)$ (J for C.M.E. Jordan).

4.2. COMMUTATION RELATIONS. First of all, we have

$$A_{\mathrm{J}} = k\langle x, y\rangle/(xy - yx - y^2);$$
$$A_{\mathrm{J}}^* = k\langle \xi, \eta\rangle/(\xi^2, \eta^2 + \xi\eta, \xi\eta + \eta\xi).$$

One first checks that the normally ordered monomials form bases of A_{J} and A_{J}^* as in Section 3. Now let the matrix $Z = \begin{pmatrix} a & b \\ c & d \end{pmatrix}$ act upon $\begin{pmatrix} x \\ y \end{pmatrix}$ and $\begin{pmatrix} \xi \\ \eta \end{pmatrix}$ preserving the defining relations. We then get the following list of relations defining the matrix quantum space $M_{\mathrm{J}}(2)$:

$$\begin{aligned}
&[a,b] - b^2 = 0;\\
&[b,c] - [a,d] + bd + db = 0;\\
&[c,d] - d^2 + ad - bc - bd = 0;\\
&[a,c] - c^2 = 0;\\
&[b,c] + [a,d] - cd - dc = 0;\\
&[b,d] - d^2 + ad - cb - cd = 0.
\end{aligned} \tag{4.1}$$

The first three relations define the endomorphism space of A_{J}^*; the second three, of A_{J}; they can also be obtained by imposing the first three relations on the coefficients of the transposed matrix.

4.3. DETERMINANT. $\mathrm{DET}_{\mathrm{J}} = D$ is defined by the action of Z upon $\xi\eta$; a straightforward calculation shows that

$$\mathrm{DET}_{\mathrm{J}} = ad - b(c+d) = da - (c+d)b. \tag{4.2}$$

It certainly is not central, because it anticommutes with b and c.

4.4. ADJUGATE MATRICES. Left and right adjugate matrices of Z exist but do not coincide:

$$\begin{pmatrix} a & b \\ c & d \end{pmatrix}\begin{pmatrix} d-b & -b \\ a+b-c-d & a+b \end{pmatrix} = \begin{pmatrix} d+b & -c+d-a+b \\ -b & a-b \end{pmatrix}\begin{pmatrix} a & b \\ c & d \end{pmatrix} = \begin{pmatrix} D & o \\ O & D \end{pmatrix}.$$

The following result proved by Korensky (Moscow University) shows that $M_q(2)$ is fairly large, but its Hopf envelope coincides with the automorphism scheme of A_J, that is, the one-dimensional torus.

4.5. PROPOSITION. *(a) The following monomes form a k-basis of $M_J(2)$:*

$$d^\delta b^\beta a^\alpha, d^\delta c^\gamma a^\alpha, Db, D^k d^\varepsilon,$$

where $\alpha, \beta, \delta \geq 0, \gamma, k \geq 1, \varepsilon = 0, 1$.

(b) Let $h : M_J(2) \to GL_J(2)$ be the Hopf envelope of $M_J(2)$. Then $h(b) = h(c) = 0$, $h(a) = h(d)$.

Sketch of proof. Korensky first represents A_J as $k\langle a, b, c, d, D\rangle$ factorized by the relations (4.1) and (4.2) and then applies the reduction technique explained in [Be] to this situation, ordering the generators by $D < d < b < c < a$. A long but straightforward application of the diamond lemma furnishes the basis of reduced monomials described in the first part of the proposition.

On the way, one gets a number of relations among which are $D^2 b = 0$, $Dc = Db$, $Dd^2 = D^2$. The first two show that $h(b) = h(c) = 0$ because $h(D)$ is invertible, being a multiplicative element of a Hopf algebra. Then from the third one, it follows that $h[(a-d)d] = 0$, hence $h(a) = h(d)$, since $h(d)$ is invertible.

A moral of this example seems to be that regularity alone is not sufficient for a quantum space to have an interesting quantum automorphism group. Some additional properties are needed to ensure a nontrivial quantum symmetry.

4.6. FROBENIUS ALGEBRAS. Before turning to larger regular algebras, we shall discuss the general technique of constructing quantum determinants and adjugate matrices. As was explained in Section 8 of [Ma2], they arise in a Hopf algebra coacting upon a Frobenius algebra. A graded algebra as in Section 4.1 is called a Frobenius one of dimension d, if $\dim(A_d) = 1$; $A_i = 0$ for $i > d$; and the multiplication map $m : A_j \otimes A_{d-j} \to A_d$ is a nondegenerate pairing for all $j \geq 0$.

Frobenius algebras, in particular, experimentally appear as $\mathrm{Ext}^{\cdot}_A(k \cdot k)$ for regular algebras A, but I do not know whether such an Ext-algebra is necessarily Frobenius (the nondegeneracy of multiplication is unclear, even if A is quadratic).

Let A be Frobenius. Choose a basis $\{x_i\}$ in A_1 and a right orthogonal basis $\{y_k\}$ in A_{d-1}, in the sense that $x_i y_k = \delta_{ik} a$, where $a \in A_d \setminus \{0\}$. Let E be a bialgebra coacting upon A (linearly upon A_1). If Z (resp. V) is the multiplicative matrix corresponding to $\{x_i\}$ (resp. $\{y_k\}$) and $\mathrm{DET}(Z) = D$ is defined by the coaction on A_d, we have $ZV^t = DI$. Using a left orthogonal basis in A_{d-1}, we get a relation $WZ^t = DI$ in the same way.

The following proposition may be used to show the centrality of the determinant.

4.7. PROPOSITION. *Assume that there exists an automorphism $\tau : H \to H$ such that $\tau(Z) = Z^t$ and $\tau(D) = D$. Then D commutes with all coefficients of Z iff*

$$Z(\tau(W) - V^t)Z = 0.$$

If, in addition, H is a Hopf algebra, this condition reduces to

$$\tau(W) = V^t.$$

Proof. From $WZ^t = DI$, we get $\tau(W)Z = \tau(D)I$ so that $\tau(W)$ and V^t are two adjugate matrices for Z. Elements of Z commute with D iff Z commutes with DI, that is,

$$Z\tau(W)Z = ZDI = DIZ = ZV^tZ.$$

EXAMPLES. (a) Consider the commutation relations (3.1)–(3.7) for the format (0,0) and the exceptional value of $q : q^2 = -1$. They define a bialgebra $\hat{M}_i(2)$ whose k-basis is formed by monomials $a^\alpha b^\beta (cb)^\varepsilon c^\gamma d^\delta$; $\alpha, \beta, \gamma, \delta, \varepsilon \geq 0$. We have

$$\det(Z) = ad + ibc \neq \det(Z^t) = ad + icb,$$

and both elements are noncentral.

(b) In general, consider the bialgebra E_q constructed in Section 3.1 for a format $(0, \ldots, 0)$ and the arbitrary values of deformation parameters q_{ij}. For $s = [j_1, \ldots, j_n] \in S_n$, put

$$q_s = q_{[j_1,\ldots,j_n]} = \prod_{\substack{a<b \\ j_a > j_b}} q_{j_a j_b}$$

and

$$P(q) = \sum_s q_s^2.$$

Demidov proved that $\mathrm{DET}(Z) = \mathrm{DET}(Z^t)$ if $P(q) \neq 0$. Moreover, define for $k \in \{1, \ldots, n\}$ a polynomial $P_k(q)$ by the following formula. Let S_{n-1}

be the permutation group of $\{1,\ldots,n\}\backslash\{k\}$. Then

$$\sum_{s\in S_{n-1}} \operatorname{sgn}(s)q_s\xi^{s^{-1}(1)}\ldots\xi^{s^{-1}(n)} = P_k(q)\xi^1\ldots\hat{\xi}^k\ldots\xi^n.$$

Demidov proved that if $P_k(q)\neq 0$ for all k, then the equality $v^t = \tau(W)$ is true if the following relations are satisfied: $P_k(q)q_{k1}\cdots q_{kk}q_{k+1,k}\cdots q_{nk}$ *is independent of* k. In particular, if $q_{ij} = q$ for all i,j, then $\mathrm{DET}(Z)$ is central, with possible exception of the case when q^2 is a root of unity of degree $\leq (n-1)$.

4.8. QUADRATIC FROBENIUS ALGEBRAS. Since the Frobenius property is so important for the existence of quantum determinants, quantum Cramer identities, and so on, it is important to understand their structure.

For a quadratic algebra A, we can define several canonical Koszul complexes. First of all, in $A^*\otimes A$, there is the Casimir element $\Sigma\xi^i\otimes x_i$ used in Section 3.6, with square zero. We can multiply by it from the right or left and dualize.

Assume first that A^* is Frobenius of dimension d. Consider the complex of left-graded A-modules

$${}_AK^i = A\otimes A_i^*, \delta = \text{right multiplication by } \Sigma x_i\otimes\xi^i$$

and the complex of right-graded A-modules

$$K_{A,i} = A_{d-i}^*\otimes A, \delta = \text{left multiplication by } \Sigma\xi^i\otimes x_i.$$

Both have nontrivial (co)homology at least at one place: $k\otimes A_d^*$ is a space of nonbounding cocycles in ${}_AK^d$, and similarly in $K_{A,0}$.

4.9. THEOREM. *Assume that A^* is a quadratic Frobenius algebra and that ${}_AK^\cdot$ and $K_{A,\cdot}$ are acyclic outside of trivial places. Then A is regular iff all roots of the Poincaré polynomial $\sum\dim(A_i^*)t^i = P_{A^*}(t)$ are roots of unity.*

Proof. Two Koszul complexes ${}_AK^\cdot$ and $K_{A,\cdot}$ are A-dual resolutions of k (or, rather, A_d^*), where the duality is induced by the multiplication $A_i^*\times A_{d-i}^*\to A_d^*$. Moreover, we have $P_{A^*}(t)P_A(-t^{-1}) = t^d$. Therefore, the roots-of-unity condition is necessary and sufficient for the polynomial growth of A.

If one does not assume that A^* is Frobenius, it is necessary to construct the A-dual complex to ${}_AK^\cdot$ directly. Denote it by $L_{A,\cdot}$.

One easily shows that the following two classes of algebras coincide:

(i) Regular quadratic algebras A such that ${}_Ak$ has a resolution of the type

$$\cdots \to V_i \otimes A(-i) \to \cdots \to A \xrightarrow{\varepsilon} {}_Ak \to 0.$$

(ii) Quadratic algebras A, for which ${}_AK^{\cdot}$ and $L_{A,\cdot}$ are acyclic except for trivial places, and roots of $P_{A^*}(t)$ are roots of unity.

The acyclicity of $L_{A,\cdot}$ was intensively studied (cf. [BacF]). Algebras with this property are called Koszul algebras, and their category has remarkable stability and duality properties.

Not all Koszul algebras are regular; on the other hand, not all regular algebras are quadratic: In [ArTdvB], a family of three-dimensional regular algebras defined by cubic relations was constructed. Both classes seem to be very interesting from the viewpoint of quantum groups.

4.10. FURTHER DEFORMATIONS OF GL(2). This subsection was added after very stimulating discussions with John Tate, Mike Artin, and Bill Schelter at the University of Texas at Austin in October 1989. First, these discussions clarified the reason for the smallness of $M_J(2)$ and $GL_J(2)$ after localizing with respect to DET: We tried to combine relations for $\begin{pmatrix} a & b \\ c & d \end{pmatrix}$ and $\begin{pmatrix} a & b \\ c & d \end{pmatrix}^t$, but the transposition essentially kills even the usual algebraic automorphism group of A_J (cf. Section 4.2)!

A successful way of adding missing relations was suggested by Artin. If one introduces a parameter q into the relation, say, $yx - qxy + x^2$, then it defines the space $A^{2|0}_{q'}$ for $q \neq 1$ and A_J for $q = 1$. Hence, one should apply a limiting procedure.

The net result of it is the following picture. Define four coordinate quantum spaces:

$$A : k\langle x,y\rangle/(yx - xy - y^2);$$
$$A^* : k\langle \xi,\eta\rangle/(\xi^2, \eta\xi + \xi\eta, \eta^2 + \eta\xi);$$
$$B : k\langle X,Y\rangle/(YX - XY + X^2);$$
$$B^* : k\langle \Xi,H\rangle/\langle H^2, H\Xi + \Xi H, H\Xi - \Xi^2\rangle.$$

Define the matrix quantum space $M = k\langle a,b,c,d\rangle/(R)$ by the condition that the following maps extend to coactions:

$$\begin{pmatrix} x \\ y \end{pmatrix} \to \begin{pmatrix} a & b \\ c & d \end{pmatrix} \otimes \begin{pmatrix} x \\ y \end{pmatrix}; \qquad \begin{pmatrix} \Xi \\ H \end{pmatrix} \to \begin{pmatrix} a & b \\ c & d \end{pmatrix} \otimes \begin{pmatrix} \Xi \\ H \end{pmatrix};$$
$$(X,Y) \to (X,Y) \otimes \begin{pmatrix} a & b \\ c & d \end{pmatrix}; \qquad (\xi,\eta) \to (\xi,\eta) \otimes \begin{pmatrix} a & b \\ c & d \end{pmatrix}.$$

Then M is regular of dimension 4. In particular, $\dim M_n$ is the same as that of algebras of commutative polynomials in four variables.

The determinant simplifies to

$$\mathrm{DET}\begin{pmatrix} a & b \\ c & d \end{pmatrix} = ba - ab + a^2.$$

Another interesting phenomenon discovered by Artin, Tate, and Schelter is the existence of two-parametric large $\mathrm{GL}_{p,q}(2)$. Namely, with the usual notation $A_q^{2|0} = k\langle x, y\rangle/(yx - qxy)$, we ask that

$$\begin{pmatrix} x \\ y \end{pmatrix} \to \begin{pmatrix} a & b \\ c & d \end{pmatrix} \otimes \begin{pmatrix} x \\ y \end{pmatrix}, \quad \text{resp.} (x', y) \to (x', y') \otimes \begin{pmatrix} a & b \\ c & d \end{pmatrix}$$

simultaneously define the coactions upon $A^{2|0}$ and $A_p^{2|0}$, respectively.

This leads to the commutation relations

$$ba = pab, \qquad dc = pcd, \qquad ca = qac, \qquad bd = qbd,$$
$$pcb = qbc, \qquad da - ad = (p - q^{-1})bc.$$

The determinant can be defined by

$$D = da - qbc = da - pcb = ad - q^{-1}cb = ad - p^{-1}bc.$$

We have $\Delta(D) = D \otimes D$. It is not central but normalizing:

$$[D, a] = [D, d] = 0; \qquad pDb = qbD; \qquad qDc = pcD.$$

The corresponding matrix algebra is again regular of dimension 4.[1]

5. $\mathrm{GL}_q(n)$ at the Roots of Unity: Frobenius at Characteristic Zero and the Hopf Fundamental Group

5.1. THEOREM. *Let* $1 = 1 \bmod (2)$ *and* q *be a primitive root of unity of* degree ℓ. Let $Z = (z_i^j)$ be the matrix of the fundamental corepresentation of $\mathrm{GL}_q(n)$, $Z^{(\ell)} = \left((z_i^j)^\ell\right)$. Then

(a) $[z_r^s, (z_i^j)^\ell] = 0$ for all r, s, i, j.

(b) $\Delta(Z^{(\ell)}) = Z^{(\ell)} \otimes Z^{(\ell)}$.

(c) $\mathrm{DET}_q(Z)^\ell = \det(Z^{(\ell)})$.

In other words, in the language of noncommutative geometry, we have an exact sequence of quantum groups

$$1 \to H_q \to \mathrm{GL}_q(n) \xrightarrow{\Phi_\ell} \mathrm{GL}_1(n) = \mathrm{GL}(n) \to 1,$$

[1]I heartily thank Artin, Tate, and Schelter for their permission to publish these results here.

where Φ_ℓ is a "Frobenius map in characteristic zero," $\Phi_\ell^*(Z_1) = Z_q^{(\ell)}$, and H_q is a finite quantum group whose function algebra is generated by (ζ_i^j) with the same commutation relations as z_i^j and additional ones $(\zeta_i^j)^\ell = \delta_i^j$.

REMARK. It is tempting to look at H_q as a quotient of a certain profinite "fundamental quantum group" of the usual algebraic affine group $\mathrm{GL}(n) = \mathrm{GL}_1(n)$. Unfortunately, it is unknown (to me) how to connect H_q's for different q's.

Lusztig [Lu2,3] constructs analogous of H_q's (or, rather, dual Hopf algebras) for other types of quantum groups.

Our presentation here is due to Parshall and Wang.

5.2. *Proof.* We first recall how Gauss binomial coefficients appear in noncommutative geometry: If $xy = q^{-1}yx$, the

$$(x+y)^\ell = \sum_{i=0}^{\ell} \begin{bmatrix} \ell \\ i \end{bmatrix}_q x^i y^{\ell-i},$$

where

$$\begin{bmatrix} \ell \\ i \end{bmatrix}_q = \frac{[i+1]_q \dots [\ell]_q}{[1]_q \dots [i]_q},$$

$$[i]_q = q^{i-1} + \dots + 1 = \frac{q^i - 1}{q-1}.$$

As a corollary, we obtain that, if $q^\ell = 1$ and q is primitive, then

$$(5.1) \qquad (x+y)^\ell = x^\ell + y^\ell,$$

and if, in addition, ℓ is odd, then

$$(5.2) \qquad (xy)^\ell = x^\ell y^\ell.$$

Induction by n shows that if $x_i x_j = q^{-1} x_j x_i$ for $i < j$, we have also

$$(5.3) \qquad (x_1 + \dots + x_n)^\ell = x_1^\ell + \dots x_n^\ell.$$

Now

$$\Delta(Z^{(\ell)})_i^k = (\Delta(Z)_i^k)^\ell = (\sum_j z_i^j \otimes z_j^k)^\ell = \sum_j (z_i^k)^\ell \otimes (z_j^k)^\ell$$
$$= Z^{(\ell)} \otimes Z^{(\ell)}$$

in view of Eq. (5.3), because for $j < m$,

$$(z_i^j \otimes z_j^k)(z_i^m \otimes z_m^k) = q^{-2}(z_i^m \otimes z_m^k)(z_i^j \otimes z_j^k),$$

and q^{-2} is again a primitive root of unity of degree ℓ. This proves part (b) of the theorem.

To treat part (a), it suffices to look at the $n = 2$ case, i.e., put $Z = \begin{pmatrix} a & b \\ c & d \end{pmatrix}$. Clearly, b^ℓ and c^ℓ commute with a, b, c, d. For a^ℓ, it suffices to check that $a^\ell d = da^\ell$. Put $\delta = ad - q^{-1}bc = da - qbc$. Since this determinant is central, we have $a^\ell d = a^{\ell-1}(\delta + q^{-1}bc) = \delta a^{\ell-1} + q^{1-2\ell}bca^{\ell-1} = (\delta + qbc)a^{\ell-1} = da^\ell$. Similarly, d^ℓ is central.

We leave part (c) to the reader.

6. Quantum Tori and Quantum Theta-Functions

6.1. QUANTUM TORI. Although we restricted ourselves by the consideration of quantum versions of GL and by explaining one method of construction, the reader will find many more quantum analogs of linear algebraic groups in the literature (cf. [ReTF], [Dr1], [Ta], and the references therein). In particular, quantum versions of O and Sp are known, and they give rise to two more series of GL, denoted in [Ta] by GL^{O}, GL^{Sp}, respectively, because O_q and Sp_q can be naturally realized as "closed subgroups" of these two GL's respectively, but not of the usual GL_q.

One would respect, therefore, a quantization theory for another important class of algebraic groups, namely that of abelian varieties. Here we shall take a small step in this direction, suggesting a definition of quantized theta-functions. They will be elements of the algebra of smooth functions on quantum (or "noncommutative") tori extensively studied in the noncommutative differential geometry of Connes's style.[2]

From our viewpoint, an algebraic quantum torus over a field k is defined by its function ring, which is a localization of the affine space considered in Section 3.1:

$$k\langle x_1, x_1^{-1}, \ldots, x_n, x_n^{-1}\rangle / (x_i x_j - q_{ij} x_j x_i \mid i < j).$$

Working over a topological field (like $\mathbb{C}$, $\mathbb{R}$, or $\mathbb{Q}_p$), one can extend this ring by considering Laurent series of x_i with decreasing coefficients, or various normed completions.

---

[2]See the beautiful review by Rieffel, "Noncommutative tori—a case study of noncommutative differentiable manifolds," preprint, Dept. Math., Univ. of California, Berkeley, 1989, containing a comprehensive list of references.

Although the general quantum torus is not a group, it is acted upon by k-points of the usual algebraic torus $\mathbb{G}_m^n : x_i \to t_i x_i,\ t_i \in k^*$. For $k = \mathbb{C}$ and a discrete subgroup of multiplicative "periods" $\Gamma \subset (\mathbb{C}^*)^n$, we shall try to construct almost periodic Laurent series satisfying the usual functional equation with respect to translations in Γ. They will be our quantized theta-functions.

We shall change notation to make it closer to the classical one. Essentially, we shall imagine x_k as an analog of the classical $e^{2\pi i z_k}$.

6.2. EXPONENTIAL NOTATION. (a) A new quantization parameter (former (q_{ij})) will be a skew bilinear form $\phi : \mathbb{Z}^n \times \mathbb{Z}^n \to \mathbb{R}$ (in the noncommutative differential geometry papers, 2ϕ is denoted θ, but we save θ for our theta-functions). Basic harmonics on a noncommutative torus T_ϕ will be denoted by $e_\phi(k)$, or just by $e(k)$, $k \in \mathbb{Z}^n$. They satisfy the commutation rule

$$(6.1) \qquad e(a)e(b) = e^{2\pi i \phi(a,b)} e(a+b).$$

The classical case corresponds to $\phi \equiv 0$ (or $\phi(a,b) \in \mathbb{Z}$ for all a, b). Then $e(a) = e^{2\pi i z \cdot a}$, where $z \in \mathbb{C}^n$, $z \cdot a = \Sigma z_j a_j$.

(b) Let $t \in (\mathbb{C}^*)^n = \mathrm{Hom}(\mathbb{Z}^n, \mathbb{C}^*)$. The translation by t in T_ϕ will be defined by its action t^* on the basic harmonies:

$$(6.2) \qquad t^*(e(a)) = t(a)e(a).$$

(c) The period subgroup will be defined with the help of a biadditive map $p : \mathbb{Z}^n \times \mathbb{Z}^n \to \mathbb{R}$. After exponentiation, it defines a bimultiplicative pairing $\mathbb{Z}^n \times \mathbb{Z}^n \to \mathbb{C}^*$:

$$\langle a, \tau \rangle : \tau_p(a) = e^{\pi i p(a,\tau)}.$$

All $\tau(\cdot)$, for $\tau \in \mathbb{Z}^n$, form the period subgroup Γ.

In order to ensure a functional equation of classical type and an exponential decrease rate of coefficients of our theta-functions, we shall state the following conditions. Let

$$p_s(b,\tau) = \tfrac{1}{2}(p(b,\tau) + p(\tau,b)), \quad p_a(b,\tau) = \tfrac{1}{2}(p(b,\tau) - p(\tau,b)).$$

Then we must have:

$$(6.3) \qquad p_a(b,\tau) = \phi(b,\tau),$$

$$(6.4) \qquad p_s(b,\tau) = q(b) + q(\tau) - q(b+\tau),$$

where $q : \mathbb{Z}^n \to \mathbb{C}$ is a quadratic form with $\mathrm{Im}(q) > 0$.

6.3. THEOREM. *For any $\ell : \mathbb{Z}^n \to \mathbb{C}$, consider the following noncommutative Fourier series:*

$$\theta : \sum_{a \in \mathbb{Z}^n} e^{2\pi i(q(a)+\ell(a))} e(a). \tag{6.5}$$

Then, for arbitrary $\tau \in \mathbb{Z}^n$, we have

$$\tau_p^*(\theta) = e^{-2\pi i(q(\tau)-\ell(\tau))} e(\tau)\theta. \tag{6.6}$$

Proof. In view of Eqs. (6.2) and (6.5), we have

$$\tau_p^*(\theta) = \sum_a e^{2\pi i(q(a)+\ell(a)+p(a,\tau))} e(a).$$

By replacing the summation variable a by $a + \tau$, we obtain

$$\tau_p^*(\theta) = \sum_a e^{2\pi i(q(a+\tau)+\ell(a)+\ell(\tau)+p(a+\tau,\tau))} e(a+\tau). \tag{6.7}$$

On the other hand, by using Eqs. (6.5) and (6.1), we obtain:

$$e^{-2\pi i(q(\tau)-\ell(\tau))} e(\tau)\theta = \sum_a e^{2\pi i(q(a)+\ell(a)-q(\tau)+\ell(\tau)+\phi(a,\tau))} e(a+\tau). \tag{6.8}$$

By comparing the coefficients at $e(a + \tau)$ in Eqs. (6.7) and (6.8), we see that we need the following identity,

$$q(a+\tau) + p(a+\tau,\tau) = q(a) - q(\tau) + \phi(a,\tau),$$

or, equivalently,

$$q(a+\tau) - q(a) - q(\tau) = -p(\tau,\tau) - 2q(\tau) - p(x,\tau) + \phi(x,\tau).$$

This last form immediately follows from Eqs. (6.3) and (6.4), since $p(\tau,\tau) = p_s(\tau,\tau) = -2q(\tau)$.

6.4. REMARKS AND QUESTIONS. (a) Since for $n = 1$ we have $\phi \equiv 0$, we do not obtain nontrivial quantized elliptic theta-functions in this way. All starts from dimension 2.

(b) Since $e(\tau)$ does not commute with θ, writing the functional equation (6.6) involves a choice. We have chosen to put $e(\tau)$ to the left of θ; one could write similar formulas with right automorphy factors. We already met a similar effect in the theory of supersymmetric theta-functions (cf. Chapter 2, Section 8.4).

(c) Similarly, one can construct quantized theta-functions of higher weights, theta-functions with characteristics, etc. Due to noncentrality of automorphy factors, however, a product of theta functions will, in general, not be a theta-function. However, the quotients of the type $\theta_1^{-1}\theta_2$ will be Γ-invariant, so that one can still try to construct "the field of rational functions on a quantized abelian variety."

(d) Can one define quantized jacobians and quantized moduli spaces?

BIBLIOGRAPHY

[A] E. Abe, Hopf algebras, Cambridge Tracts in Mathematics *74*, Cambridge Univ. Press (1980)

[AlGS1] L. Alvarez-Gaumé, C. Gomez and G. Sierra, Hidden quantum symmetries in rational conformal field theories, Preprint CERN-TH-5129/88.

[AlGS2] L. Alvarez-Gaumé, C. Gomez and G. Sierra, Duality and quantum groups, Preprint CERN-TH-5369/89.

[ArS] M. Artin and W. Schelter, Graded algebras of global dimension 3, Advances in Mathematics *66* (1987), 171–216.

[ArTvdB] M. Artin, J. Tate and M. van den Bergh, Some algebras associated to automorphisms of elliptic curves, Preprint, Harvard University (1988).

[BacF] J. Backelin and R. Fröberg, Koszul algebras, Veronese subrings and rings with linear resolutions, Revue Roumaine de Math. Pures et Appl. *30* (1985), 85–97.

[BarFS] M. Baranov, I. Frolov and A. Schwarz, Geometry of two-dimensional superconformal field theories (in Russian), Teor. Mat. Fiz. *16* (1985), 202–297.

[BarMFS] M. Baranov, Yu. Manin, I. Frolov and A. Schwarz, A superanalog of the Selberg trace formula and multiloop contributions for fermionic strings, Comm. Math. Phys. *111* (1987), 373–392.

[Barr] M. Barr, $*$-Autonomous categories, Lecture Notes in Mathematics *752* (1979), Springer, Berlin–Heidelberg–New York.

[BDr1] A. Belavin and V. Drinfeld, On the solutions of the classical Yang–Baxter equations for simple Lie algebras (in Russian), Funkc. Analiz *16* (1982), 1–29.

[BDr2] A. Belavin and V. Drinfeld. On the classical Yang–Baxter equation for simple Lie algebras (in Russian), Funkc. Analiz. *17* (1983), 69–70.

[Be] G. Bergman, The diamond lemma for ring theory, Advances in Mathematics *29* (1978), 178–218.

[BGG] I. Bernstein, I. Gelfand and S. Gelfand, Schubert cells and cohomology of spaces G/P (in Russian). Uspekhi Mat. Nauk. *28* (1973), 3–26.

[BS] A. Beilinson and V. Schechtman, Determinant bundles and Virasoro algebras, Comm. Math. Phys. *118* (1988), 651–690.

[C1] A. Connes, Sur la théorie noncommutative de l'intégration, Lecture Notes in Mathematics *725* (1979), Springer, Berlin–New York.

[C2] A. Connes, Introduction to noncommutative differential geometry, Lecture Notes in Mathematics *1111* (1984), 3–16, Springer, Berlin-Heidelberg–New York–Tokyo.

[C3] A. Connes, Noncommutative differential geometry, Publ. Math. IHES *62* (1984), 257–360.

[C4] A. Connes, The von Neumann algebra of a foliation, Lecture Notes in Physics *80* (1979), 145–171, Springer, Berlin–New York.

[Ca] P. Cartier, Homologie cyclique, Sém. Bourbaki, Exp. *621* (1984).

[CFQS] J. Cohn, D. Friedan, Z. Qiu and S. Shenker, Covariant quantization of supersymmetric string theories, Preprint EFI 85-90, (1985), Chicago University.

[ChaP1] V. Chari and A. Pressley, Fundamental representations of Yangians, Preprint Tata (1988).

[ChaP2] V. Chari and A. Pressley, Representations of Yangians, Preprint Tata (1988).

[ChaP3] V. Chari and A. Pressley, Quantum R-matrices and intertwining operators, Preprint Tata (1988).

[Che1] I. Cherednik, On a method of construction of factorizable S-matrices in elementary functions (in Russian), Teor. Mat. Fiz. *43* (1980), 117–119.

[Che2] I. Cherednik, Quantum groups as hidden symmetries of classic representation theory, Preprint, Moscow University (1988).

[Che3] I. Cherednik, On "quantum" deformations of irreducible finite-dimensional representations of gl_N, Sov. Math. Dokl. *33* (1986), 507–510.

[Che4] I. Cherednik, On irreducible representations of elliptic quantum R-algebras, Sov. Math. Dokl. *34* (1987), 446–450.

[D1] P. Deligne, Letter to Yu. Manin, Sept. 25 (1987).

[D2] P. Deligne, Catégories tannakiennes, Preprint IAS (1988).

[D3] P. Deligne, Le groupe fondamental de la droite projective moins trois points, Preprint IAS (1988).

[De] M. Demazure, Désingularization des variétés de Schubert généralisées, Ann. Sci. ENS *7* (1974), 53–88.

[DM] P. Deligne and J. Milne, Tannakian categories, Lecture Notes in Mathematics *900* (1982), 101–128, Springer, Berlin–Heidelberg–New York.

[Dr1] V. Drinfeld, Quantum groups, Proc. Int. Congr. Math. Berkeley 1986, Vol. 1, 798–820.

[Dr2] V. Drinfeld, Quasi-Hopf algebras (in Russian), Preprint, FTINT, Kharkov (1989).

[Dr3] V. Drinfeld, Hamiltonian structures on Lie groups, Lie bialgebras and a geometrical interpretation of the classical Yang–Baxter equations (in Russian), Doklady *268* (1982), 285–287.

[Dr4] V. Drinfeld, Hopf algebras and the quantum Yang–Baxter equation (in Russian), Doklady *283* (1985), 1060–1064.

[Du] M. Dubois-Violette, On the theory of quantum groups, Preprint Orsay-LPTHE 89/20 (1989).

[FeTs1] B. Feigin and B. Tsygan, Additive K-theory, Lecture Notes in Mathematics *1289* (1987), 67–209, Springer, Berlin.

[FeTs2] B. Feigin and B. Tsygan, Cyclic homology of algebras with quadratic relations, universal enveloping algebras and group algebras, *ibid.*, 210–239.

[FrJ] I. Frenkel and N. Jing, Vertex representations of quantum affine algebras, Preprint (1988).

[FT] L. Faddeev and L. Takhtajan, *Hamiltonian approach to solitons theory*, Springer, Berlin–Heidelberg–New York (1983).

[G1] D. Gurevich, Yang–Baxter equation and a generalization of the formal Lie theory (in Russian), Doklady *288* (1986), 797–801.

[G2] D. Gurevich, Quantum Yang–Baxter equation and a generalization of the formal Lie theory, Reports Dept. of Math., Stockholm University, Stockholm, Vol. 24 (1986), 33–123.

[G3] D. Gurevich, Trace and determinant in algebras connected with Yang–Baxter equations (in Russian), Funkc. Analiz. *21* (1987), 79–80.

[GeS] M. Gerstenhaber and S. Schack, Quantum groups as deformations of Hopf algebras, Preprint, University of Pennsylvania at Philadelphia and State University of New York at Buffalo (1988).

[Ha] T. Hayashi, q-analogues of Clifford and Weyl algebras. Spinor and oscillator representations of quantum enveloping algebras, Preprint, Nagoya University (1988).

[He] D. Hejhal, On Schottky and Teichmüller spaces, Advances in Mathematics *15* (1975), 133–156.

[J1] M. Jimbo, A q-difference analogue of $U(g)$ and the Yang–Baxter equation, Lett. Math. Phys. *10* (1985), 63–65.

[J2] M. Jimbo, A q-analog of $U(\mathrm{gl}(N+1))$, Hecke algebra and the Yang–Baxter equation, Lett. Math. Phys. *11* (1986), 247–252.

[Jo] V. Jones, A polynomial invariant of knots via von Neumann algebras, Bull. AMS *12* (1985), 103–111.

[K1] V. Kac, Lie superalgebras, Advances in Mathematics *26* (1977), 8–96.

[K2] V. Kac, Representations of classical Lie superalgebras, Lecture Notes in Mathematics *676* (1978), 597–626, Springer, Berlin–Heidelberg–New York.

[Ka1] M. Karoubi, Connexions, courbures et classes charactéristiques en K-théorie algébrique, Proc. Canadian Math. Soc. *2* (1982), 19–27.

[Ka2] M. Karoubi, Homologie cyclique et K-théorie algébrique, I, II, C. R. Acad. Sci. Paris *297* Série I (1983), 447–450, 513–516.

[Kas] Ch. Kassel, Le résidu non commutatif (d'après M. Wodzicki), Sém. Bourbaki, Exp. *708* (1989).

[Ko] T. Kohno, Monodromy representations of braid groups and Yang–Baxter equations, Ann. Inst. Fourier *37* (1987), 139–160.

[Kst] D. Kastler, *Cyclic cohomology within the differential envelope*, Hermann, Paris (1988).

[L1] V. Lyubashenko, Hopf algebras and vector symmetries (in Russian), Uspekhi *41* (1986), 185–186.

[L2] V. Lyubashenko, Superanalysis and solutions of the triangle equations (in Russian), Ph.D. thesis, Kiev University (1986).

[LBR] C. LeBrun and M. Rothstein, Moduli of super Riemann surfaces, Comm. Math. Phys., *117* (1988), 159–176.

[Le] A. Levin, Supersymmetric algebraic curves, Ph.D. Thesis, Moscow University (1988).

[Lö] C. Löfvall, On the subalgebra generated by one-dimensional elements in the Yoneda Ext-algebra, Lecture Notes in Mathematics *1183* (1986), 291–338, Springer, Berlin–Heidelberg–New York.

[LQ] J. Loday and D. Quillen, Cyclic homology and the Lie algebra homology of matrices, Comm. Math. Helv. *59* (1984), 565–591.

[Lu1] G. Lusztig, Quantum deformations of certain simple modules over enveloping algebras, Advances in Mathematics *70* (1988), 237–249.

[Lu2] G. Lusztig, Modular representations and quantum groups, Preprint, Harvard University (1988).

[Lu3] G. Lusztig, Quantum groups at root of 1, Preprint, Harvard University (1989).

[Ma1] Yu. I. Manin, *Gauge field theory and complex geometry*, Springer, Berlin–Heidelberg–New York–London–Paris–Tokyo (1988).

[Ma2] Yu. Manin, *Quantum groups and noncommutative geometry*, Montreal University (1988).

[Ma3] Yu. Manin, Neveu–Schwarz sheaves and differential equations for Mumford superforms, Journal of Geometry and Physics, *5* No. 2 (1988), 161–181.

[Ma4] Yu. Manin, Some remarks on Koszul algebras and quantum groups, Ann. Inst. Fourier *37* (1987), 191–205.

[Ma5] Yu. Manin, Multiparametric quantum deformation of the general linear supergroup, Comm. Math. Phys. *123* (1989), 123–135.

[MaD] Yu. Manin and V. Drinfeld, Periods of p-adic Schottky groups. Journ. f. d. Reine u. Angew. Math. *262/263* (1973), 239–247.

[MaR] Yu. Manin and A. Radul, A supersymmetric extension of the Kadomtsev–Petviashvili hierarchy, Comm. Math. Phys. *98* (1985), 65–77.

[McL] S. MacLane, Natural associativity and commutativity, Rice University Studies *69* (1963), 28–46.

[MMNNU] T. Masuda, K. Mimachi, Y. Nakagami, M. Noumi and K. Ueno, Representations of the quantum group $SU_q(2)$ and the little q-Jacobi polynomials, Preprint (1988).

[Mu] D. Mumford, *Lecture on curves on an algebraic surface*, Princeton, Princeton University Press (1966).

[NYM] M. Noumi, H. Yamada and K. Mimachi, Zonal spherical functions for the quantum homogeneous space $SU_q(n+1)/SU_q(n)$, Preprint (1988).

[OF1] A. Odesskii and B. Feigin, Sklyanin's algebras associated with elliptic curves (in Russian), Preprint (1988).

[OF2] A. Odesskii and B. Feigin. Elliptic Sklyanin's algebras (in Russian), Preprint (1989).

[PaS] V. Pasquier and H. Saleur, Symmetries of the XXY-chain and quantum $SU(2)$, Preprint Saclay SPhT/88–187 (1988).

[PS] I. Penkov and I. Skornyakov, Cohomologie des $\mathcal{D}$-modules tordus typiques sur les supervariétés des drapeaux, C. R. Acad. Sci. Paris *299* Série I (1985), 1005–1008.

[Pe] I. Penkov, Borel–Weyl–Bott theory for the classical Lie supergroups (in Russian), Itogi Nauki i Tehniki *32* (1988), 71–124, VINITI, Moscow.

[Pr] S. Priddy, Koszul resolutions, Trans. AMS *152* (1970), 39–60.

[Ra] A. Radul. The Schwarz derivative and the Bott cocycle for Lie superalgebras of string theories (in Russian), Numerical Methods for Differential Equations, Moscow, MGU (1986), 53–67.

[Re1] N. Reshetikhin, Quantized universal enveloping algebras, the Yang–Baxter equation and invariants of links, I, II. LOMI preprints E-4-87, E-17-87, Leningrad (1988).

[Re2] N. Reshetikhin, Quasitriangle Hopf algebras, solutions of Yang–Baxter equations and links invariants (in Russian), Algebra i Analiz *1* (1989).

[ReTF] N. Reshetikhin, L. Takhtajan and L. Faddeev, Quantization of Lie groups and Lie algebras (in Russian), Algebra i Analiz *1* (1989), 178–206.

[Ro] A. Rosenberg, Noncommutative affine semischemes and schemes, Reports Dept. of Math., Univ. Stockholm (1988).

[Ros1] M. Rosso, Finite-dimensional representations of the quantum analog of the enveloping algebra of a complex simple Lie algebra, Comm. Math. Phys. *117* (1988), 581–593.

[Ros2] M. Rosso, An analog of P.B.W. theorem and the universal R-matrix for $U_h sl(n+1)$, Preprint, Centre Mathématique, École Polytechnique, Palaiseau (1988).

[Ros3] M. Rosso, Analogues de la form de Killing et du théorème d'Harish-Chandra pour les groupes quantiques, Preprint, Centre Mathématique, École Polytechnique, Palaiseau (1989).

[S] R. Saavedra, Catégories Tannakiennes, Lecture Notes in Mathematics *265* (1972), Springer, Berlin–Heidelberg.

[Sk1] E. Sklyanin, On algebraic structures connected with Yang–Baxter equations (in Russian), Funkc. Analiz. *16* (1982), 22–34.

[Sk2] E. Sklyanin, *ibid.* II, *ibid.* *17* (1983), 34–48.

[SkTF] E. Sklyanin, L. Takhtajan, L. Faddeev, Quantum inverse scattering transform method (in Russian), Teor. Math. Fiz. *40*, (1979), 194–220.

[Sm] D.-J. Smit, The quantum group structure in a class of $d = 2$ conformal field theories, Preprint THU-88/33, Univ. of Utrecht, 1988.

[T] B. Tsygan, Homology of matrix Lie algebras over rings and Hochschild homology (in Russian), Uspekhi *38* (1983), 217–218.

[Ta] M. Takeuchi, Quantum orthogonal and symplectic groups and their embedding into quantum GL, Preprint (1988).

[V] A. Vaintrob, Deformations of complex superspaces and coherent sheaves on them (in Russian), Itogi Nauki i Tehniki *32* (1988), 125–211, VINITI, Moscow.

[VM] A. Voronov and Yu. Manin. Supercell decompositions of flag superspaces (in Russian), Itogi Nauki i Tehniki *32* (1988), 27–70, VINITI, Moscow.

[Vo] A. Voronov, Relative position of the Schubert supervarieties and their desingularization (in Russian), Funkc. Analiz i ego pril. *21* (1987), 72–73.

[VS] L. Vaksman and Ya. Soibelman, Algebra of functions on the quantum group $SU(2)$ (in Russian), Funkc. Analiz. *22* (1988), 1–14.

[W1] S. Woronowicz, Twisted $SU(2)$-group. An example of a noncommutative differential calculus, RIMS, Kyoto Univ. *23* (1987), 117–181.

[W2] S. Woronowicz, Compact matrix pseudogroups, Comm. Math. Phys. *111* (1987), 613–665.

[W3] S. Woronowicz, Tannaka–Krein duality for compact matrix pseudogroups. Twisted $SU(N)$-groups. Preprint, Univ. Leuven (1988).

[Wo1] M. Wodzicki, Noncommutative residue, Chapter 1. Fundamentals, Lecture Notes in Mathematics *1289* (1987), 320–399, Springer, Berlin–Heidelberg–New York.

[Wo2] M. Wodzicki, Cyclic homology of differential operators. Duke Math. Journ. *54* (1987), 641–647.

[Wo3] M. Wodzicki, Excision in cyclic homology and in rational algebraic K-theory, Ann. of Math. *129* (1989), 591–639.

INDEX

(Note: II.4.1 refers to Chapter 2, Section 4, Subsection 1)

www.ingramcontent.com/pod-product-compliance
Ingram Content Group UK Ltd.
Pitfield, Milton Keynes, MK11 3LW, UK
UKHW011100140125
453571UK00007B/192

9 780691 607160